the PLANT DOCTOR

the PLANT DOCTOR

a practical guide to having a healthy garden

SONIA DAY

KEY PORTER BOOKS

Copyright © 2006 Sonia Day

All rights reserved. No part of this work covered by the copyrights hereon may be reproduced or used in any form or by any means—graphic, electronic or mechanical, including photocopying, recording, taping or information storage and retrieval systems—without the prior written permission of the publisher, or, in case of photocopying or other reprographic copying, a licence from Access Copyright, the Canadian Copyright Licensing Agency, One Yonge Street, Suite 1900, Toronto, Ontario, M6B 3A9.

Library and Archives Canada Cataloguing in Publication

Day, Sonia
    The plant doctor : a practical guide to having a healthy garden / Sonia Day.
Includes index.

ISBN 1-55263-737-9

1. Gardening—Canada. 2. Garden pests. 3. Plant diseases.
I. Title.

SB453.D38 2006   635'.0971   C2005-906146-4

The publisher gratefully acknowledges the support of the Canada Council for the Arts and the Ontario Arts Council for its publishing program. We acknowledge the support of the Government of Ontario through the Ontario Media Development Corporation's Ontario Book Initiative.

We acknowledge the financial support of the Government of Canada through the Book Publishing Industry Development Program (BPIDP) for our publishing activities.

Key Porter Books Limited
Six Adelaide Street East, Tenth Floor
Toronto, Ontario
Canada M5C 1H6

www.keyporter.com

Design and layout: Ingrid Paulson

Printed and bound in Canada

06 07 08 09 10 6 5 4 3 2 1

This book is dedicated to all the people who have admired beautiful gardens and wondered if they could create one themselves

Contents

Introduction 9

CHAPTER 1 STARTING OUT 13
Know Your Hardiness Zone 13
Sun or Shade? It's Crucial Information.. 18
How Much Sunlight Do Plants Need?... 20
What to Do about Dry Shade 21
Why Some Plants Won't Grow under a Black Walnut Tree 24

CHAPTER 2 BUYING PLANTS 27
How to Shop at a Garden Center.. 27
What Kind of Plant to Buy: Annual or Perennial?..................................... 30
The Different Types of Bulbs 31
Buying by Mail Order 34
Be Wary of These Invasive Plants 36

CHAPTER 3 SOIL BASICS 41
How to Tell What Kind of Soil You Have 42
Understanding Soil pH.. 43
Ways to Improve Soil 44
What Fertilizers Can and Can't Do... 49
Which Is Best: Organic or Chemical?... 53
Some Plants That Benefit from Chemical Fertilizers 54
Tips about Some Popular Plant Care Products 55
The Magic of Manure Tea 57

CHAPTER 4 PLANTING 59
Where to Plant 59
When to Do It 61
If Jack Frost Makes a Return Visit 62
How to Plant Most Things... 64
Important Planting Tips 65
Planting under Trees... 67
The Problem of Wind.. 68
Different Bulbs Need Different Treatment 69
Planting in Containers... 72

  Planting Roses in Cold Climates..74
  The Marvel of Mulch 75

CHAPTER 5 **ONGOING PLANT CARE**.. 81
  Take a Tour of Your Garden.. 81
  The Touchy Issue of Watering... 83
  Weeding: The Inescapable Task 85
  Getting Rid of Weeds.. 87
  Deadheading... 92
  Shearing 94
  Staking 96
  Pinching... 99
  Thinning 100
  Dividing Plants That Have Grown Too Big................................. 100
  Volunteers: A Mixed Blessing in Many Gardens 103
  What to Do When Winter Comes 106
  When Spring Rolls around Again.. 110
  When Roses Won't Produce Flowers 110

CHAPTER 6 **PROTECTING PLANTS FROM PROBLEMS** 113
  How to Cultivate Good Gardening Habits 113
  What IPM Means and How It Affects You 116
  Beneficial Bugs in the Garden 117
  Does Companion Planting Work?... 119
  The Different Kinds of Pesticides Sold at Garden Centers 121
  Chemical Pesticides: Know What You're Using........................... 123
  Organic Pesticides Can Be Toxic Too...126
  Common Garden Pests and How to Combat Them 134
  The Four Kinds of Plant Diseases.. 149
  Common Diseases That Affect Plants 151
  Ways to Keep Critters Away.. 156

 Acknowledgments 162
 Photo Credits.. 162
 Bibliography... 163
 Index.. 165

Introduction

Beautiful gardens don't just happen. TV makeover shows like to foster the illusion that it can all be done in a weekend, and that by simply picking up the phone, we can order "a garden" in much the same way that we can order a pizza. However, the reality is very different. Working with nature takes time. Lots of time. Creating a garden is a slow process, not a quick fix. You can't simply plunk a bunch of plants in the ground, then walk away and expect them to fend for themselves.

Choosing the right kind of plants, and figuring out how treat them, can actually be a bit bewildering. One reason is the list of choices, which keeps getting longer. There's also no shortage of experts telling us what to buy. But gardening fads come and go, and something that's in this year may be out the next. That makes decisions difficult. Which plants will suit our garden? Where's the best place to put them? When should we plant and how? Are there things we need to do to our soil before we go shopping at the garden center? How should we care for our plants once they are established? And what's the best course of action when things go wrong?

This book answers those basic questions and more. You won't find any gushing descriptions of plants in the following pages, but you will find lots of practical information about making wise choices and then helping those purchases thrive. There are chapters on the importance of knowing the kind of climate you garden in and on assessing your growing conditions. You'll learn about ways to improve the soil and get plants off to a good start, then care for them once they're established. The pros and cons of chemical and

Previous page: All plants need care. A daily check is the best way to keep them healthy and beautiful. These geraniums are flourishing because the gardener waters them regularly and removes spent blooms.

organic fertilizers and pesticides (an often neglected topic) are extensively covered, and there are detailed descriptions of pests and diseases that you're likely to encounter, as well as practical suggestions on getting rid of them.

If all this sounds like work, it is. Undeniably. Gardening isn't for the indolent. However, as any keen gardener will tell you, being around plants is a very

pleasurable kind of work. It's soothing, satisfying, and rewarding—all at the same time. The extraordinary Dutch painter Vincent Van Gogh (who loved gardening) discovered that. He once wrote, in a letter to his brother Theo: "Just for one's health ... it is very necessary to work in the garden every day and see the flowers growing." Using this book as a guide, you may wind up feeling the same way.

Left: Maintaining a garden like this is undeniably lots of work. But it's also very satisfying. Many keen gardeners find that taking a daily tour in their gardening grubbies is the highpoint of their day.

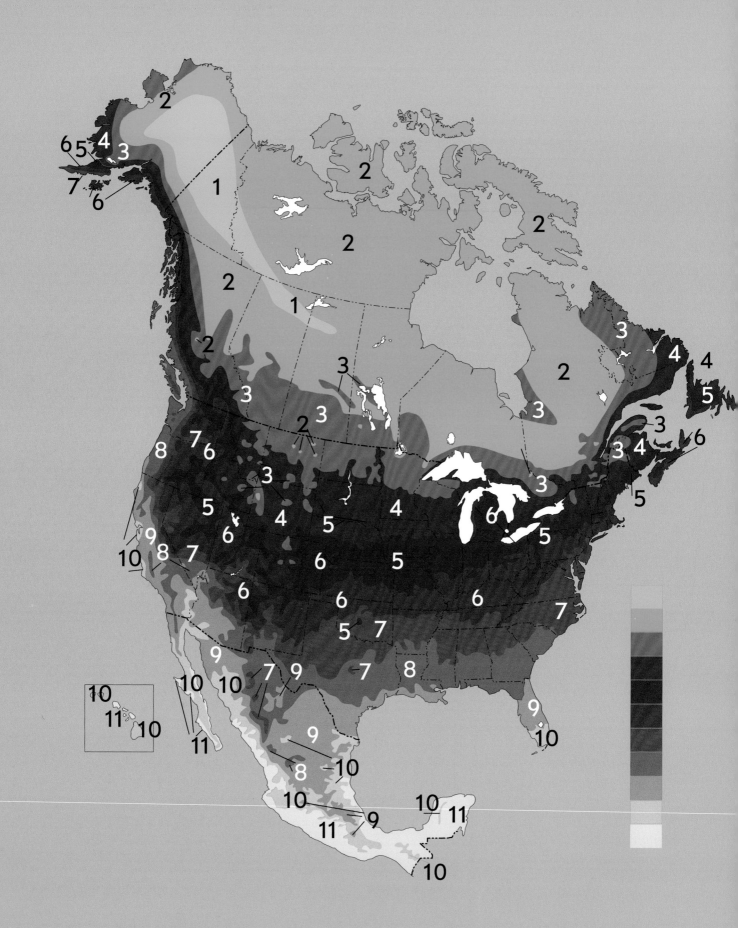

Starting Out

KNOW YOUR HARDINESS ZONE

Many experienced gardeners can rattle off their hardiness zone number quicker than their birth dates—and with good reason. The more you grow things, the more your zone number becomes burned into your brain because it's one key to establishing a healthy, trouble-free garden.

What is a hardiness zone? For the purpose of agriculture and horticulture, all the regions of North America are divided up into zones. The main factor in delineating these zones is how low the temperature drops in winter because that determines whether or not a plant will survive until next spring. But other factors, such as the length of the growing season, soil conditions, and local fluctuating temperatures come into play too.

Generally speaking, the warmer the weather, the higher the zone number. For example, Zone 1 is way up in the Arctic, but by the time you travel south to Zone 11, you're down Mexico way.

It's important to know your hardiness zone number. Without it, you can wind up wasting money on purchases that are too tender to survive the winter in your area. Consult the hardiness zone map in this book, but don't trust it implicitly. It's only a guide. For clarification, ask at a garden center or find out from a neighbor who gardens because various local climatic influences can skew the numbers. (U.S. and Canadian hardiness zones are somewhat interchangeable, but there are a few differences. See the bibliography at

Left: Zone maps like this one are a useful guide. But, for clarification, check at a garden center or ask a neighbor, because local climatic conditions can skew the zone numbers.

This little city garden sits in a microclimate. Sheltered by trellises, shrubs, and trees, it experiences warmer temperatures than nearby gardens that have no protective surroundings.

the back of this book for more information.)

Bear in mind, too, that while garden center staff certainly know what the local hardiness zone is (and can make suggestions about what to grow), they usually don't restrict themselves to selling things that are winter hardy in your zone. Much nursery stock—plants, shrubs, and trees—is raised in warm regions like California, then shipped, ready-potted, all over North America, including northern regions. Even if some of those offerings on display won't survive local winters, garden centers tend to sell them anyway.

Always check plant tags stuck into pots. The hardiness zone should be listed on the tag. If it isn't, ask. This number should be the same number as the zone you live in. It can also be *lower* down the numerical scale, but not higher. For instance, you can safely plant something that's marked "Hardy to Zone 4" if you live in Zone 6. If it's tough enough to survive in the frigid –30°F (–31.7°C) minimum winter temperatures of Zone 4, it's not going shrivel up and die on you in your relatively balmy −10°F (−20°C) minimum winter temperatures. However, you wouldn't select a plant that's "Hardy to Zone 8" because that plant can cope only with a minimum winter temperature of 10° to 20°F (−9.5° to −3.9°C), which is much warmer than the winters where you live.

Microclimates and How They Work

All kinds of conditions can make the temperature in your garden warmer or cooler than the mean temperature in your hardiness zone. Which way the garden faces obviously plays a role. A southern aspect is warmer than a northern one. But a nearby body of water, such as a pond or lake, may have a warming influence too. So, too, can concrete or brick walls, fences, hedges, overhanging vines, rows of trees, or any sort of enclosure because those external factors protect the growing area and make it warmer. Conversely, if you grow things in an exposed bare backyard without any trees, fences, or buildings surrounding it, or if you garden high off the ground—on a rooftop or balcony, for instance—you may notice temperatures that are persistently much cooler than those experienced by other local gardeners. These situations are called microclimates and it's important to be aware of them because they may make it necessary to ratchet your own hardiness zone up or down a notch or two.

With a high-rise balcony or rooftop, always select plants that hardy to *two zones colder* than your local zone. Research has shown that it's invariably colder and tougher for plants to survive up in the sky. However, in a protected garden at ground level in the same city as that high-rise, you may be able to get away with planting things that are designated for zones far warmer than yours. As well, there may be microclimates with wildly differing temperatures in different areas of the same garden. (See "Understanding Frost Pockets," page 16.)

Zonal Denial: A Risky Concept

With global warming making winter temperatures more unpredictable, many gardeners have begun to embrace "zonal denial." That is, they're tossing the concept of hardiness zones on the compost heap, planting whatever they please, then mulching those plants heavily before the winter and keeping their fingers crossed. This can certainly work, particularly if you garden in a large city. That's because urban high-rises tend to generate a lot of heat, which sticks around, making the ground and indeed the whole area warmer than it is outside the city. There are some surprising examples of shrubs, plants, and trees surviving in unexpected locations. A *Cercis canadensis* 'Forest Pansy,' traditionally regarded as hardy only to Zone 6, may come through a very cold winter in Zone 5 unscathed if it's planted in a protected location.

There's no question that zonal denial can be tremendously exciting. Many gardeners find it a challenge to get "difficult" things to grow. However, ignoring hardiness zones is a risky practice, and not for everyone. If you're putting in expensive perennials for the first time, why take the chance? It's wise to pick the right plants for where you live because discovering that they have died during the winter can be very discouraging, and it's costly to replace them.

Understanding Frost Pockets

Frost is a deposit of ice crystals. These form when the temperature at ground level falls low enough to turn water into ice. Curiously, the air temperature doesn't have to be freezing for frost to form. It can be as high as 39°F (4°C) in the air, but if the ground temperature dips to 32°F (0°C), tiny droplets of water start crystallizing into ice as they descend.

At night, the ground temperature is almost always cooler than the air temperature because heat accumulated in the soil radiates back into the atmosphere as the night progresses. This erosion of heat continues until

Frost pockets tend to form at the bottom of slopes. When choosing plants for such sites, always stick to winter-hardy species. Do not plant vegetables there.

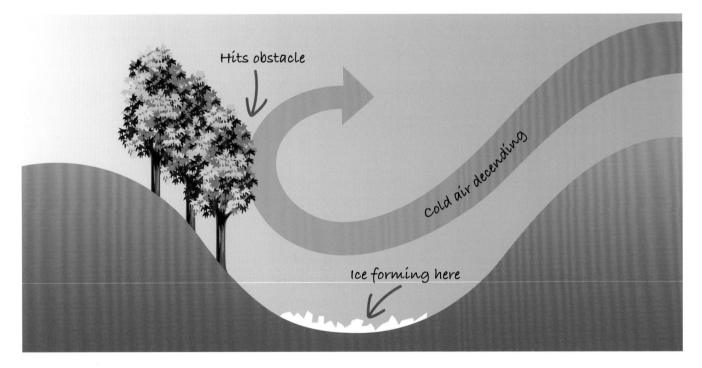

the sun returns the following morning to warm things up again. As a result, the sharpest frosts often occur right before dawn or just after when the sun is still very weak.

This means that it pays to listen very carefully to weather forecasts! When meteorologists make predictions, they are usually talking about the temperature of the air about 6 ft (1.8 m) *above* the ground, not on the ground itself. Many gardeners have had the experience of going to bed, presuming that there won't be any frost. Then they've headed outside the next morning and found, to their horror, that their tender plants all got zapped by Jack Frost during the night.

Whether frost injures plants or not depends upon a number of factors. Hardy types of perennials normally start arming themselves against the cold before the ground temperature dips to 32°F (0°C). Their leaves and stems undergo chemical and metabolic changes as days shorten, and tissues become "hardened" against the anticipated frost. A few species transfer water to tissues where it can freeze harmlessly. However, tender plants don't have that resistant capacity. That's why leaves and stems of annuals turn black or go mushy after a heavy frost.

Native plants are nice, but...

Some purists advocate sticking to native plants only (that is, plants that are indigenous to your area) because in doing so you create a more "ecologically natural" garden and those choices will certainly cope with local temperatures. But why be so restrictive? And how do we determine exactly what *is* native anyway? So many plants have found their way to North America from so many different climates and regions of the world. By all means, seek out the native kinds, but be adventurous too. You'll find some wonderful plants out there, waiting to be discovered. Just check that they're winter hardy in your area.

Many gardens contain "frost pockets" where freezing temperatures seem to hit far faster than elsewhere in the garden. This usually happens in hollows or at the bottom of a hill because cold air is denser than warm air and, like water, it flows downhill. However, frost pockets can also occur on flat ground, where cold air swirls up against an obstacle—such as a hedge or fence—and then settles and turns into ice because it can't escape.

When choosing plants for a known frost pocket, it is a good idea to stick to winter-hardy species that can cope with sudden dips in temperature. And avoid creating a vegetable garden in a frost pocket.

SUN OR SHADE? IT'S CRUCIAL INFORMATION

It's a good idea to wait a year before installing a new garden (or renovating an old one). Yes, an entire year. While it can be tempting to rush out, after moving into a new place, and buy a lot of new plants, we can wind up making the wrong kinds of purchases because we have only a vague idea about our growing environment.

What deceives most people at the beginning is the amount of sunshine that our gardens get. Often, it's a lot less than we think. This is particularly true of small urban and suburban properties surrounded by other homes. It's easy to presume that a front or backyard gets plenty of sun all day because it faces east, or south, or west. But only as our first summer in that home wears on does it become evident that the house next door completely blocks the sunlight in the afternoon, or that the tree in the backyard leafs out in June, casting heavy shade, or that our own garage prevents much light from getting to anything planted beside it.

It's crucial to know how much sun different areas of your garden receive because when plant labels say "Full sun," they generally mean six to eight hours of *unobstructed* sunshine a day. Many popular flowering perennials, annuals, and spring bulbs need that amount, and if you have less, you'll have to settle for plants that tolerate some shade—and maybe a lot of it. However, this is no reason for gloom. The choice of plants for shady gardens is improving all the time and in our era of global warming, it can be a boon rather than a disadvantage for them to be shielded from broiling hot sun all day.

Any of these signs may mean that a plant is not getting enough sunlight for its particular needs:

- The plant is becoming tall and leggy.
- It hasn't reached its potential size.
- Leaves are sparse, sickly looking, or falling off. They may also be changing shape and becoming more elongated and/or a darker green.
- It's not flowering, or flowers are small and few and far between.

How to Determine the Level of Sunlight in Your Garden

Go outside. Stand in the front, the back, and at the sides of your property. Observe the position of the sun carefully, and where—and when—the sunlight hits the ground throughout the day. Take note of obstructions that cast shadows, reducing the amount of light in various areas of the garden. These obstructions can be adjacent buildings, fences, trees, hedges, shrubs, tall plants, even your neighbor's mini-van, which is always parked in a certain spot, shielding your flower bed. Check the level of sunlight in all seasons of the year, not just spring when you're eager to plant something.

What's important to bear in mind—along with the physical obstructions—is that in northern climates, the sun shifts its position drastically throughout the summer. By June, it rises much farther north than it does in March. Then by fall, it moves lower in the sky again. This means that an area of the garden that's in shade much of the day during spring may receive several hours of direct sunshine by July because the sun is higher at noon. But by September, with the sun moving lower again and casting longer shadows, the same area will become shady once more, a phenomenon that plays an important role when choosing plants. Such a location would not be a good choice, for instance, for sun-loving perennials that bloom late in the summer, like black-eyed Susans *Rudbeckia fulgida*.

The fact is, in modern cities and suburbs where people live in homes packed closely together, most gardens receive a lot less direct sunlight than we think.

The Different Kinds of Shade You May Have
- *Light shade* means plants get from four to six hours of direct sunshine a day, or they're positioned in a spot that gets low levels of shade all day.
- *Partial shade* means plants get two to four hours of direct sunlight or dappled shade all day (see below).
- *Full shade* means the plants get only reflected, indirect light.
- *Dense shade* means there's very little light reaching plants at all.

What Is "Dappled" Shade?
This term is used a lot by garden centers and it often puzzles people. It simply means shade from surrounding trees. Buildings cast solid shade because sunlight can't penetrate through them. But some light filters through trees, particularly deciduous ones. This may happen early in the season when they're leafing out, late in the year when leaves fall, and on windy days throughout the growing season when leaves are swaying to and fro.

Some plants that don't like solid shade will perform well if shade is dappled

This public park receives a lot of dappled shade cast by tall trees. In such situations, pick plants such as impatiens (above) that can cope with a mix of light.

because a bit of sunlight keeps penetrating through the foliage. But others—particularly low-light houseplants that originated in shady, damp tropical jungles—hate this kind of shade. (Never put a peace lily *Spathiphyllum wallissii* that's been indoors all winter outside under a tree on a sunny, windy day. Those "sunspots" getting through the leaf canopy when the wind blows can actually burn holes in a peace lily's leaves. Full shade is the right outdoor location for this kind of plant.)

HOW MUCH SUNLIGHT DO PLANTS NEED?

While it varies enormously, the inescapable truth is that most plants need some light, either direct or indirect, to thrive, and that in deep, dense, all-day shade, the choices are much more limited. Look at the labels stuck into pots at garden centers. Most carry symbols that will tell you what the growing requirements are. Where sun versus shade is concerned, growers divide plants into three basic groups:

- A completely white circle means the plant needs full sun—that is, at least six hours of sunshine every day (with no shade from trees or surrounding buildings getting in the way).

- A circle split into two—one half white, the other black—means part sun is required. This is at least half a day—three hours of sunshine. Most plants seem to prefer morning sun. If you're planting something in a west-facing location (and the sun doesn't arrive until after midday), the plant will probably need at least four hours of sun to flourish, but it may also do well in partial shade all day.

- A solid black circle means the plant will thrive with no direct sunshine at all, but this doesn't mean "total darkness"—simply that it will thrive in an area that gets full or dense shade all day. Often, it also means that this kind of plant dislikes direct sunshine falling on its leaves.

What if a plant carries a tag with two symbols—"full sun" and "part sun," for instance? This denotes a plant that is happy in a range of conditions. It may thrive in full sun, but will do equally well in partial shade. Or it's a shade-loving plant that won't throw a hissy fit if exposed to sun.

Always take the time to read labels stuck in plant pots. They will usually tell you if the plants are a suitable choice for your garden.

WHAT TO DO ABOUT DRY SHADE

Areas underneath trees are bound to be damp and cool—and full of plants that are lush and green, right? Yes and no. While this is certainly true of some forests, tropical jungles, and the "bosky glades" described in fairy tales, the reality is quite different in most urban gardens. If you have large, mature trees, you're probably stuck with the challenging phenomenon known as "dry shade." Homeowners are often shocked to discover that the soil under their trees stays as dry as a bone all year and is very poor quality. They also discover that it's impossible to grow anything under the tree canopy because the tree roots take up all the moisture and crowd out everything else.

Foam flower *Tiarella cordifolia* is a pretty native plant that survives well in dry shade.

Dry shade is in fact very common, and it doesn't happen just under trees. This annoying condition also can be found in soil next to adjoining buildings. In these instances, there's often the additional obstacle of mortar from the wall leaching into the soil, increasing its alkalinity. If nothing seems to grow well next to a wall, test the pH and adjust it, if necessary. (See page 43.)

Many people try—in vain—to coax plants and lawns to flourish in dry shade. But before planting anything, or laying down grass seed or sod, it's imperative to improve the soil first. It's also a good idea to mulch plants every spring, work new organic matter into a lawn, then keep up a watering regime all summer, in order for those underplantings to survive. In some instances, where it's impossible to do much to the soil, growing things in raised planters that sit on the top of tree roots may be the only practical option.

See page 67 for more on how to plant under trees.

Perennial Plants That Can Usually Cope with Dry Shade

- *Convallaria majalis*, lily of the valley*
- *Dodecatheon*, Shooting Star**
- *Epimedium rubra*, barrenwort**
- *Galium odoratum*, sweet woodruff*
- *Geranium macrorrhizum*, bigroot geranium**
- *Lamium maculatum*, deadnettle
- *Tiarella cordifolia*, foam flower**
- *Vinca minor*, periwinkle*

- *Viola odorata,* common violets*

The following perennial plants may also do well, but are more picky about soil conditions and/or the amount of shade cast:

- *Alchemilla mollis,* lady's mantle**
- *Asarum canadense,* wild ginger***
- *Bergenia*
- *Brunnera macrophylla***
- Bulbous plants such as snowdrops and dwarf narcissus (humus levels in the soil should be high)
- Hostas**
- *Pachysandra terminalis,* Japanese spurge
- *Polygonatum,* Solomon's seal**
- *Pulmonaria,* lungwort**

Do not, under any circumstances, plant common goutweed *Aegopodium podagraria* (also known as bishop's weed). Although it flourishes in shade (and indeed anywhere) this darn plant will keep spreading ferociously and its roots are virtually impossible to remove once they've been introduced. Another variety, *A. variegata,* which has pale green leaves outlined with cream, is less invasive but, even so, is not recommended for small gardens.

Annual Plants That May Survive in Dry Shade

- *Coleus:* A foliage plant with colorful leaves, available in many varieties. Purple kinds do well in shade.
- *Impatiens:* Ho-hum leaves, but small pink, orange, red, or white flowers and a reliable bloomer; easy to grow.

Coleus come in many colorful varieties. They may do well under trees, provided they are watered regularly.

---

* Caution: Invasive plants, which may prove too much of a good thing in a small garden.
** Recommended
*** Recommended, but avoid in areas where raccoons are a problem, as they dig it up.

- *Tuberous begonias:* Dark green leaves, large flowers in vivid yellow, red, orange, and pink

Buy the above as started plants and fertilize them regularly, or they won't produce many flowers. (See page 59 for more on planting.)

If you want to plant a tree, then grow things underneath it, try a Russian olive *Eleagnus angustifolia*. It's handsome, has silvery leaves, will tolerate dry sites, and doesn't seem to mind a bit of company gathering at its base. However, it tends to self-seed everywhere, and you will constantly need to yank out the seedlings.

WHY SOME PLANTS WON'T GROW UNDER A BLACK WALNUT TREE

This is one of the most commonly asked questions on gardening hotlines across North America. People call up complaining that their vegetable plants "look funny" or that their perennials and shrubs are wizened and not blooming when planted in the vicinity of the notorious native black walnut *Juglans nigra*.

The culprit is a chemical called juglone, which is in every part of this tree — branches, bark, leaves and roots. The toxic effects may spread as wide as the tree's canopy, which in bigger specimens will reach 50 ft (15 m) or more. However, juglone is most bothersome to anything planted directly around the base of the tree. What's puzzling, however, is that there seem to be no hard and fast rules about which plants are affected. Some gardeners report that vegetable plants, particularly members of the cabbage family, are most bothered by juglone even if they're planted far from the tree. Others find all kinds of perennials refuse to grow properly. Still others insist that black walnut's difficult reputation is exaggerated. One fact is hard to dispute: black walnut leaves do have a toxic effect on turf, causing bare or shrivelled patches, so do not let leaves lie around on the lawn after they've fallen off the tree. Don't compost the leaves either, as juglone remains in anything it has touched, and its toxicity does not decline over time.

Although often touted as a good tree for home gardens because it's a North American native species, black walnut has other drawbacks too. When grown in poor soil, it tends to be messy (branches break off frequently in winds). It also produces an abundance of big, heavy, green fruits, which encase the seeds (i.e., nuts). These fall continuously from midsummer onwards, and are inclined to bonk unwary gardeners on the head. They also attract squirrels, and although the nut meats certainly taste good, their casings are so hard that you need a carpenter's vise and the strength of Hercules to crack them open.

On the plus side, the leaf canopy of black walnut is high and does not cast dense shade, so some plants will thrive underneath it. Tough perennials are the best choices. But all things considered, enjoy black walnuts in the wild or public parks. Do not introduce a new specimen to a small garden or plant one where it is likely to prove problematic to your neighbors.

The following plants are most likely to cope with the rigors of growing under or near a black walnut:

- Astilbes
- Buxus, boxwoods
- Campanulas
- Clematis
- Daphne
- Daylilies
- *Hibiscus syriacus*, rose of Sharon
- Hostas
- *Monarda didyma*, bee balm
- Peonies
- Sedums
- Tulips (and other kinds of spring-flowering bulbs)
- Viburnums

ALERT!
Why plants lose their variegated look

When a plant is "variegated," it means it produces leaves that have two colors on the same leaf—two shades of green, one light, one dark, for instance. Or green outlined with red. Or cream in the center, with a green outline around the edge.

If the plant stops producing these variegated leaves and reverts to solid green only, the problem can usually be traced to insufficient sunlight or, less frequently, infertile soil. Reversion often afflicts variegated shrubs, such as Spireas, *Weigela*s and *Euonymus*, and the annual foliage plant, coleus.

Whatever the reason, cut off the offending plain-colored leaves, removing the entire stem, the moment you notice them. The entire plant may quickly revert to this single color if you wait too long.

And if the problem persists, and you want to keep the plant's variegated appearance, move it to a sunnier location in the garden.

Buying Plants

2

Start out with healthy plants and they'll be much easier to grow. You can now shop for good specimens in many different locations: garden center chains, owner-operated nurseries large and small, plant specialists, big box stores, supermarkets, neighborhood greengrocers, garden decor stores, horticultural groups' plant sales, mail-order companies, you name it. However, as with everything else we buy, some sources are better than others.

HOW TO SHOP AT A GARDEN CENTER

- Patronize an established business that sells reliable plant stock that's been properly nurtured, is disease and pest free, and has been correctly labeled.

- Look for evidence that the garden center has a qualified horticulturist on staff. Usually you'll see a logo displayed at the entrance that shows an affiliation with a horticultural organization. These vary from state to state. In Canada, a triangular logo with CCHT means at least one employee is a Canadian certified horticultural technician. You can consult these very knowledgeable folks for advice. (Be warned: Garden centers at big box stores often depend upon inexperienced summer help, usually students, who don't know a hosta from a heuchera.)

- Avoid "fly-by-night" operations set up at curbside in the summertime. Their plant offerings often come from questionable sources, and may

Left: It's worth building a relationship with a local garden center. They're the best source of advice on what works in your area. But avoid shopping on spring weekends, when lineups are longer. Weekday evenings are better.

harbor bugs and diseases. Then when you go back to complain, they've already packed up and left.

- Bigger isn't necessarily better. A compact plant with plenty of healthy-looking green buds is almost always preferable to a tall, leggy one that has already leafed out and is in flower. Smaller plants may in fact be a more economical buy than big ones because they generally go through less transplant shock and get established quicker.

- Examine plants carefully before buying. Avoid anything with leaves that are yellowing, drooping, curling up, going brown at the tips, or falling off.

- Check for insect damage (holes in the leaves, for example) and signs of disease (such as moldy spots on leaves and stems).

- If you do notice insects after getting home with your purchases, spray the plants with a mixture of 1 cup (250 mL) of insecticidal soap combined with about 1 tbsp (15 mL) of rubbing alcohol. Leave plants in their containers for a few days before planting. Inform the place where you bought the plants (and if the infestation is serious, take the plants back).

- Diseased-looking plants are best put in a sealed plastic bag immediately and put in the garbage, not the compost heap. If you return diseased plants to the garden center, make sure they stay in the bag.

- If plants are sitting on racks in hot sunshine and look completely dried out, don't buy them.

- Pick up pots and cell paks and examine their undersides. If roots are poking out of the drainage holes, the plants have been potted for too long and they may not transplant successfully into your garden. Poke a finger into the soil on top. It shouldn't feel dried out, baked hard, or compacted.

ALERT!
Don't buy too much

It's easy to go overboard at garden centers, particularly on a warm weekend in spring, but remember that you can always go back for more. If you take on too much planting at once, the job will seem overwhelming. There's the temptation to "get it over" and cram plants into the wrong place, or leave them lying around, still in their pots, until they dry out, die, and have to be thrown out.

- Many better garden centers provide a guarantee with expensive perennials (also trees and shrubs). This is a piece of paper that you receive at the cash desk, saying you can return the plant if it dies within two years. Keep this guarantee in a safe place (but remember that it's incumbent on you to care for the plant properly).

- Read labels carefully for information on hardiness and growing requirements. If labels are missing, ask somebody. (It's surprising how often shoppers switch or steal labels.)

- Shop early in the season for best availability and quality. In most areas of North America, this is March, April, and/or early May.

- Avoid weekend shopping if you can. The moment the weather warms up, garden centers get jam packed, particularly on Saturday and Sunday afternoons. On weekdays, you're more likely to find staff willing to answer questions. Shopping at the supper hour during the week—from 6 p.m. onwards—is also a good idea.

- Before going to a garden center, do your homework. Read some gardening books, catalogs, and magazines. Make a list. Take along pictures of plants you like. Figure out whether your garden gets sun or shade or a mixture of the two. Assess what kind of soil you have and what you think you'd enjoy growing. Garden center staff hate customers who come in without any plan or ideas. "Don't just tell us that you want some plants for your garden," says one garden center owner. "We need some information on the type of garden you have in order to help you."

- Garden centers are more environmentally conscious than they used to be, and many will recycle old plastic pots. But don't simply haul your empties along and dump them in the parking lot. Some pots are usable. Others aren't. It's courteous to ask store staff first.

- It's worth establishing a relationship with small, neighborhood garden centers. These are often a bit more expensive than big box stores, but when something goes wrong, you can usually consult the owners for advice. They are always knowledgeable about the growing conditions in your area and can actually save you money.

- Plant sales held by horticultural societies, gardening-oriented organizations, and community groups are often sources of good, cheap plants. They hold these sales in spring and sometimes fall. Watch local newspapers and community bulletin boards for announcements.

WHAT KIND OF PLANT TO BUY: ANNUAL OR PERENNIAL?

Perennials

Perennials stay in the garden year round. They are sold mostly in plastic pots of varying sizes. The most common is a 4-in (10-cm) pot (also confusingly called a 1 qt, in some locations.) You'll also see 8-in (20-cm or 1-gal) pots and 16-in (41- cm or 2-gal) pots. Larger, established perennials are often expensive and usually aren't worth the extra outlay because they will take longer to settle in than smaller versions. (The exception is ornamental grasses, which take ages to get established. Buy the biggest containers of grasses that you can afford.)

Getting a bargain

July is a surprisingly good time to purchase some perennials. The spring rush is over, garden centers want to get rid of their excess stock, and there are real bargains out there. Some savvy gardeners wait till July before buying. However, the selection is usually limited and plants may be in rough shape from sitting too long in their containers. Be careful what you buy. Tough, easy-care perennials are the most likely to cope with being planted during the height of the summer. Get them into the ground promptly.

However, if you are planting perennials in fall, it can be worth buying bigger plants because those with extensive root systems tend to withstand frost heave better than smaller ones. (See page 62.)

Perennials sold by mail-order plant suppliers are usually shipped bare root in early spring (or fall), which means they'll arrive in cardboard boxes with shredded newspapers, sawdust shavings, or plastic "baggies" of soil swathed around them. Their roots may look a bit dismal on arrival because the plants have been kept in cold storage, but relax—they'll perk up once planted. Don't keep such plants sitting around for too long, however. If you aren't able to get them into the ground within a few days, toss some soil or leaves over the roots, water well, and leave in a shady spot until ready to plant. It can also be a good idea to soak really dry roots in lukewarm water before planting.

Annuals

Annuals are planted for one season only because they aren't hardy enough to remain in the garden over the winter months. However, what makes a plant annual rather than perennial depends upon where you live. (In warmer regions, some plants that northern gardeners regard as annuals can in fact be treated as perennials.) Annuals are sold mostly in little plastic containers called cell paks, with six (or four) compartments to a pak. They are cheaper than perennials because they last for only one season in the garden. However, increasingly, growers are also making some annuals available in larger plastic pots, but as with perennials, bigger pots mean higher prices. If you're planting a new, empty garden flower bed with popular annuals such as pansies, impatiens, and petunias, buy the cell paks because you'll need a lot of plants and it's not worth spending the extra money for larger versions. (They won't take long to get established and fill the bed.)

Larger started annuals are more useful when gardening in containers because they usually give instant results, and you need less of them in a container than in a garden.

THE DIFFERENT TYPES OF BULBS

Garden centers are making an increasing array of so-called "bulbs" available in response to gardeners' demand for new and exciting things to grow. Bulbs come from many countries and climatic conditions, and understanding the differences between them (and the right times to buy and plant bulbs) can be perplexing. They are divided into two basic groups:

> ## ALERT!
> ### Buy one, not lots
>
> Garden designers and garden centers are fond of insisting that it's a good idea to buy at least three of each plant because that's the key to making an attractive grouping in the garden. (Odd numbers are considered more aesthetically pleasing than plants grouped in twos and fours.) However, if you're trying an expensive perennial for the first time, ignore that advice. Buy only one. Why shell out lots of cash before you know for sure whether the plant will thrive in your garden? If the first one works, you can always go back for more.

- **Spring-flowering bulbs**, such as tulips, daffodils, hyacinths, anemones, crocus, alliums, and *Fritillaria*. These are the kinds most of us are familiar with. They are winter hardy and must be planted in fall. They usually go on sale in late August or September, and should be put into the ground before winter freeze-up.

- **Tropical bulbs**, often described as "tender bulbs" by gardening magazines and nurseries. These include *Agapanthus*, begonias, dahlias, gladioli, cannas, calla lilies and *oxalis*. They cannot be exposed to frost, must be planted in spring when the danger of freezing is over, and then dug up and brought back indoors in the fall. They go on sale in spring.

Also part of the latter category are amaryllis, which in northern climates are treated as houseplants and grown in containers indoors during the winter. They go on sale in fall and generally disappear from garden center shelves by Christmas.

Both categories of plants grow from underground storage organs. The gardening industry, for the sake of simplicity, collectively refers to them as "bulbs" and many, such as tulips, daffodils, and amaryllis, do indeed have bulbous shapes like onions. However, others do not resemble bulbs at all. Some—begonias and dahlias, for example—are tubers, and look more like small sweet potatoes. Cannas are flatter and rectangular, and are classified as rhizomes. A few, such as anemones, are tiny, hard corms, no bigger than a thumbnail. All have their own particular characteristics and require different treatment. (See page 69 for more information on planting bulbs.)

The vast majority of bulbs are now grown in countries such as the Netherlands, South America, and India, then shipped, dried, to North America. They are usually very high quality because inspection procedures are rigorous. Some tips on buying them:

- In garden centers, bulbs are sold in two ways: either in open cardboard boxes, nestled among wood shavings, or in net bags. The former are preferable because it's possible to pick out the good specimens and leave the doubtful-looking ones behind. Net bags are usually a bit cheaper, but may cost more in the long run because you can't see the broken or damaged specimens lurking inside.

- Don't buy bulbs packaged in plastic bags. They turn moldy quickly and bulbs need to be exposed to air.

- Bulbs should be firm and full, not squashy or dried out, or with moldy patches. Press them with a finger. If buying an amaryllis in a box, peek under the lid to check that the bulb is in good shape.

- Avoid broken or split tulip bulbs. They won't produce flowers and aren't worth planting.

- In the case of bulbous plants—such as tulips, daffodils, hyacinths, and

amaryllis—buy the biggest bulbs you can find. They are more expensive, but the bigger the bulb, the bigger the flower.

- Don't worry if the skins of the bulbs are flaking off. It won't affect their ability to grow. In fact, it often gets them off to a quicker start.

- Be careful when handling hyacinths. They can trigger a rash similar to poison ivy in susceptible individuals. The culprit isn't (as many people think) pesticides sprayed on the bulbs, but the hyacinths' outer skins. When these tear off, they break down into minute, needle-shaped crystals of calcium oxylate. These make skin itchy and we scratch, pushing the crystals deeper, prompting a painful pink rash.

- When buying spring-flowering bulbs, select varieties that will bloom at

Bulbs come in many shapes and sizes. They may also be called tubers, rhizomes, and corms. Their planting requirements vary. Always read the instructions.

different times, so you'll have a constant parade of blooms in the garden in spring. Look for "early flowering," "mid-season flowering," and "late flowering" on bulb labels.

- In northern climates, it is a good idea to buy summer-flowering bulbs such as begonias, dahlias, and cannas as started plants. There's an increasing array of pre-started bulbs being sold in garden centres, because they're easier for gardeners to grow—and they save time. There are several kinds of begonias available. The easiest (and most popular) for gardeners are the nonstop varieties, which growers pot up in late winter, then start selling in April or May, in 4-in (10-cm) pots. If you opt to grow the tubers or rhizomes yourself, it is best to start them indoors under lights or in a very sunny window.

- Because summer-flowering bulbs must be dug up and brought indoors at the end of the summer, most are best treated as container plants.

- Mail-order companies usually sell more interesting and unusual varieties of bulbs than garden centers. Many of these establishments will also take orders months in advance. If there's a particular bulb that you want to grow, it's worth finding out if you can order it—because it's very annoying to plan on growing something, then discover that it's sold out at the garden center. (This happens frequently in fall, with trendy varieties of tulips.)

BUYING BY MAIL ORDER

This can be a real boon—or a real bust—depending upon the supplier. Many experienced gardeners get into the habit of buying by mail (or via the Internet) because it's the way to obtain the kind of unusual plants that aren't stocked by your average garden center. And while most specialist suppliers and mail-order nurseries are worth dealing with, a few aren't. Here are some tips:

- Send for lots of plant catalogs! They're usually free, they make great reading in the wintertime, and most are packed with all kinds of information. They can teach you a tremendous amount about cultivating and caring for your plants even if you don't buy a thing.

- Beware of hype. Most catalogs make exaggerated claims: "tomatoes as big as tennis balls," "huge saucer-shaped blooms," and so on. And they're all fond of posing little kids beside plants like sunflowers and pumpkins to

make what they sell look bigger. Learn to read between the lines and ignore the purple prose. (Consult a tome like the *Reader's Digest A to Z Encyclopedia of Garden Plants* to discover the true dimensions of plants and flowers.)

- Order spring plants early (some suppliers start taking orders as early as January). Specialist nurseries often have very limited stocks of unusual or rare varieties and they sell out quickly.

- Plants ordered by mail usually look shriveled and off-putting when unpacked from the box because they're often kept in cold storage for months and then shipped bare root (but they'll perk up rapidly once planted). If you're the kind of person who likes to see how a plant looks before buying it, shop at a garden center instead.

Always wear gloves when planting hyacinths or handling them at a garden center. Their outer skins can trigger a painful rash that's similar to poison ivy.

ALERT!
Plant smuggling carries stiff penalties

If you see attractive plants while traveling overseas, be careful about trying to bring them back into the U.S. or Canada. Carrying plants across the U.S.—Canadian border is also fraught with complications. Many plants will simply be confiscated outright by customs authorities. With others, you need to file phyto-sanitary certificates and other documents in advance. And with a few rare varieties of plants like orchids, you may even be arrested on the spot because it's illegal to remove those plants from their native habitats! Regulations regarding plants are always changing. Never try to smuggle them in your luggage. It's best to check with customs before you go away if there's something you think you'll want to bring back.

- Beware of flashy mail-order plant catalogs that arrive unsolicited in the mail, offering plants that sound amazingly cheap, coupled with lotteries to take part in, at no charge. Their plants will probably be undersized, overpriced, and definitely run of the mill. You're better off visiting a garden center.

- A good mail-order supplier will tell you when and how they'll send your plants, and approximately how much the shipping charges will be. Avoid those that are vague about timing. (You don't want to wait till summer for the plants to arrive.) They'll also include comprehensive planting instructions in the package.

- Cross-border buying—that is, ordering plants from Canada when you live in the U.S. and vice versa—can be a nightmare and take weeks. Customs officials on both sides of the border may require various clearance documents, depending upon the plant, and then be extremely slow about processing them, during which time your precious plant will sit in a customs shed, unwatered and drying out. You may also be stuck with stiff brokerage fees. Before placing an order abroad, always ask the supplier what you'll be required to do to bring the plant into the country (in some instances, the procedure is so complicated, it's better to find a local source for the plant) and then verify that information with U.S. or Canadian customs.

BE WARY OF THESE INVASIVE PLANTS

Invasive plants—that is, plants that rampage everywhere, crowd out everything else, and become impossible to get rid of—are now a huge problem around the world. There are hundreds of such plants in North America. Most are nonindigenous species, and while they may be well behaved in their native habitats, they can quickly turn into noxious bullies in the growing

Ribbon grass *Phalaris arundinacea* var. 'picta' (foreground) and black-eyed Susans *Rudbeckia fulgida* are invasive plants. Avoid them in small gardens.

conditions they encounter here. These plants often have beautiful flowers, foliage, or other pleasing attributes, so their use in gardens has mushroomed. But think twice before introducing invasive plants to your front or backyard, even though you'll see many of them for sale everywhere. Garden centers usually describe these plants as "vigorous" or "fast growing" but the word "invasive" is conveniently avoided.

That said, whether a plant becomes invasive or not depends greatly upon the climate and local growing conditions. Something that's a delight in a Chicago urban garden may be regarded as a horrible nightmare by people who live in the South. One example is lantana, which has become a popular container plant in northern climates. (Many gardeners put it outside in summer, then bring it indoors for the winter.) Yet in Florida (and Australia) this pretty tropical shrub with orange-yellow flowers grows year round, and is now such a nuisance weed that certain jurisdictions have resorted to strafing it with herbicide. In some situations, thuggy members of the plant world aren't all bad. Most are likely to become a nuisance only in small to medium-sized front yards and backyards. If you have lots of space, invasive plants can actually be a boon because they fill in large areas quickly, are seldom bothered by insects or diseases, and are easier to maintain than other plants so long as you keep whacking them back and digging out chunks of the roots when they spread too far. Invasive plants can also be useful as ground covers in shady spots under trees and in areas where nothing else will grow due to poor soil or other environmental problems.

Original species of plants are often the worst offenders when it comes to invasive capabilities. New hybrids developed by growers from the "momma" plant generally do not seem to be as troublesome. For example, purple coneflowers *Echinacea:* While new seedlings of standard, pink-flowered *Echinacea* will keep popping up and may become too much of a good thing, some of the new *Echinacea* hybrid varieties, with white or orange flowers, can actually be difficult to bring into flower!

Depending upon where you live and the size of your garden, you might want to steer clear of these invasive plants:

Artemisias (particularly *A. ludoviciana* and the popular new *A. Ludoviciana* 'Limelight')
Black-eyed Susans *Rudbeckia fulgida*
Boston ivy *Parthenocissus tricuspidata*
Chinese bittersweet vine *Celastrus orbiculatus*. (American bittersweet, *Celastrus scandens*, is not as invasive.)
Creeping bellflowers *Campanula rapunculoides*
Crown vetch *Coronilla varia*
Evening primrose *Oenothera*
Euphorbia s., certain kinds: *E. griffithii*, *E. myrsinites*, and *E. polychroma* may be particularly troublesome.
Dame's rocket *Hesperis matronalis*
Joe Pye weed *Eupatorium purpureum*
Lady's mantle *Alchemilla mollis*
Lysimachia s., certain kinds: Gooseneck loosestrife *L. clethroides*, golden creeping Jenny *L. nummularia* "Aurea" and whorled loosestrife *L. punctata* may be particularly troublesome.
Mints (all kinds)
Obedient plant *Physostegia*
Ornamental strawberry *Fragaria*
Japanese knotweed *Polygonum cuspidatum*
Periwinkle *Vinca minor*

Plume poppy *Macleaya cordata*
Purple coneflower *Echinacea*
Purple loosestrife *Lythrum salicaria*
Ribbon grass *Phalaris arundinacea* var. 'Picta' (also called gardener's garters, reed canary grass, and other names)
Silver lace vine *Polygonum aubertii* (also called silver fleece vine)
Sweet woodruff *Asperula odorata*
Trumpet vine *Campsis grandiflora* (also known as Chinese trumpet creeper or Trumpet honeysuckle)

Soil Basics

3

The old saying, "Tend the soil, not the plants," has become a bit of a cliché, but it's still right on the mark. To have healthy plants, you need healthy soil. Period. Too many new homeowners race out to garden centers in spring and stock up on plants without thinking beforehand about the soil they have to work with. When their purchases don't thrive, they blame the plants or the garden center. In the vast majority of instances, however, the problem can be traced to poor soil, or the wrong kind of soil. If plants start off well, then seem to struggle and lose their vitality as summer progresses, the fault probably lies down where the roots are, not with the visible growth above the surface.

Soil is basically made up of three components: rock particles (ground-up sand), fine silt particles (clay), and organic matter (such as decomposed leaves). Mix those three together in equal proportions. Toss in some minerals and living organisms like earthworms and bacteria. Stir the lot, creating lots of air pockets, and you get an ideal soil. Horticulturists call this magic mixture "loam."

Loam is well aerated, easy to work, and retains moisture and nutrients well. Unfortunately, it's also in short supply. Very few gardens are blessed with perfect loamy soil. Instead, it's either too light (so that nutrients and moisture get leached out quickly) or too heavy (so the surface becomes hard and compacted and drainage is slow).

Left: When creating beds, don't scrimp on the soil. It's a good idea to top up around plants every spring with some fresh soil or compost.

HOW TO TELL WHAT KIND OF SOIL YOU HAVE

Dig down about 9 in (23 cm) into your soil and pick up a handful. If it's dark, loose, and crumbly, yet it can still be squeezed together into a ball, count yourself lucky: that's loam. Most gardeners are more likely to have soil containing either too much sand or too much clay. Sandy soil will feel gritty and contain large particles. It falls apart easily in the palm and won't stay together in a ball. Clayey soil, however, is the opposite. It clumps easily, is heavy, and feels slick (or sticky). When wet, clay also has a slight sheen. When dry, it's hard as a brick.

Other indicators: Water drains away quickly from sandy soil, and you can usually scratch into its surface without difficulty, using a garden tool. With clay, water will remain on the surface for a while and often puddle in lower spots. Clay feels cold to the touch in spring. It's lumpy when you dig into it, and the surface develops deep cracks in hot, dry weather.

Which is worse? Sandy or clayey soil? It's a toss-up. The sandy kind is often deficient in nutrients, and plants dry out quickly because there are too many air pockets around the soil particles. With clay, the reverse happens. Nutrients are retained better in clay, but the soil may get so compacted that roots don't receive sufficient air. Drainage is also poor, so roots rot.

It can be hard to determine what kind of soil predominates in your garden. Some properties, particularly those close to ravines or rivers, may contain a mixture of both types of soil. Gardeners in these locations are often surprised to discover that they have a sandy swath sitting right next to an area of heavy clay. Or there may be a thin layer of good soil on top of compacted sandy (or clayey) underpinnings. The latter is a common phenomenon in new subdivisions, where developers have the habit of stripping off the topsoil before starting to build. The subsoil (along with construction debris) then gets tamped down hard by their bulldozers and, after the project ends, a couple of inches of soil are restored on top. The end result is a very sorry kind of growing medium. If you live in a newly constructed home, it will almost certainly be advisable to amend the soil around it.

Whatever the conditions, remember that most plants are happy in a layer of loam 12 in (30.5 cm) deep, and that few gardens deliver that kind of environment without a bit of help.

UNDERSTANDING SOIL pH

Mention pH and many gardeners immediately look apprehensive, because it sounds so scientific. But all you really need to understand is this: an important characteristic of soil is its acidity or alkalinity. That's pH. It's worth taking the trouble to determine the pH (which means potential hydrogen) of your soil and then taking steps to change the acid/alkaline balance if necessary.

Soil-testing kits are sold everywhere now. Lab tests (which require mailing samples in) are the most accurate. But a simple do-it-yourself kit (available at the hardware store or a garden center) will provide a basic idea of pH levels. Be sure to take several samples of soil from different spots in the garden.

The pH scale runs from 1.0 (most acidic) to 14.0 (most alkaline). In a typical garden, the pH may be anywhere between 4 and 9.

Most plants prefer a neutral pH—slightly below 7 on the scale. If the pH is lower than 6, nutrients such as copper, iron, manganese, and zinc can't be absorbed by most plants, so their growth is likely to be poor. The ideal catchall number cited by horticulturists is 6.5 (that is, slightly acidic).

To Raise pH (and Make Soil Less Acidic)

To raise pH, add lime, which is sold in bags at garden centers as a fine, white powder or in pellets or granules. It's best to use *dolomitic* or *ground* limestone (sometimes labeled "agricultural" or "garden" lime). Avoid hydrated, slaked or builder's lime, or quicklime as these may burn plants and damage soil. And never add lime at the same time as manure (composted or raw), as they combine to release ammonia gas, which wastes nitrogen.

Amounts to use vary, depending upon soil type. Follow instructions on the bag. As a general rule, a sandy soil will require an application of 4 oz per square yard (114 g per 0.84 square meter). while clayey soil needs more—12 oz per square yard (340 g per 0.84 square meter). Sprinkle on the ground (preferably in the fall), and water in well. Choose a windless day and wear goggles. Don't apply lime more than once per year. It is best to raise pH slowly over an extended period.

To Lower pH (Make Soil More Acidic)

To lower pH, add horticultural sulfur. This is a yellow powder (or granules) sold in bags and often labeled "flowers of sulfur." Garden centers may recommend using aluminum sulfate, which is powdered bauxite treated with sulfuric acid. (The most common use of this stuff is to acidify the soil surrounding hydrangeas and make their flowers blue.) But over time, aluminum sulfate may actually poison certain plants. Plain sulfur is preferable.

Peat moss and pine needles can also be mixed in to make soil more acidic, but vast amounts are needed to make a discernable difference to the pH. Mixing these organic materials with sulfur will work faster.

Soil test results usually indicate sulfur quantities to add. Follow instructions and handle sulfur with caution as most people are sensitive to it. Sprinkle on no more than 1 lb per 100 sq ft (.50 kg per 9 sq mts) and don't be tempted to add more to speed up the process of making soil more acidic, as it's toxic to plants when overused.

Soil next to urban sidewalks and roads is often very alkaline because limestone (an ingredient of concrete) keeps leaching into it. This can make it hard to maintain acid-loving plants like rhododendrons and heather in a front yard. If you want to grow these in such a location, it will almost certainly be necessary to lower the pH.

Remember: changes in the pH don't have to be dramatic. Even a tiny shift in the scale, up or down, can make a tremendous difference to how things grow in your soil.

WAYS TO IMPROVE SOIL

There is a variety, not several a variety of products on the market now that are designed to improve poor soil. For large gardens, ordering these amendments by the truckload from a landscaping supply company is best because it's cheaper. But for city residents, with no space for large trucks to maneuver into tiny spaces, bagged products from garden centers are more practical. Whatever the product, buy larger packages. Smaller ones invariably cost more, and once the job of creating or renovating a flower bed starts, you'll be surprised at how much you need.

Topsoil

Every garden center sells bags marked "topsoil," but quality varies enormously because contents (and their ratios) aren't normally listed on labels and there are no industry standards for selling it. Pick up the bags. If they weigh a ton, the soil probably contains a lot of clay. If they're relatively light, the principal ingredient is likely to be peat moss.

Good bagged topsoil looks dark and rich in organic matter. It's been sifted free of lumps and stones, and is not heavy with water. It may be sold in its natural form or sterile (which means the soil has been heated to a high temperature to kill disease spores, insect larvae, and weed seeds). But the watchword when buying the bagged kind is caution because not all topsoil is tops.

Add topsoil to gardens only. Do not be tempted to use it—even the sterile, sifted kind—for container gardening. Although topsoil is cheaper than specially formulated growing mixes, it is not a good environment for most potted plants because its composition is too dense and heavy.

Peat Moss

Two kinds are sold at garden centers, which can be confusing. *Sphagnum moss* is gray and stringy and used mostly in flower arranging. *Peat moss* is dry, brown, and flaky and sold in compressed, plastic-wrapped bales. The latter is the kind to add to the garden.

People often think they're adding "fertilizers" by mixing peat moss into garden soil, but has virtually no nutrient value. The chief benefit of peat is its capacity to retain water and the fact that (unlike compost) it's usually sterile and weed free. Add a bale of peat moss to a clayey flower bed to improve aeration. Conversely, if the flower bed is too sandy, peat will stop water from draining away too fast.

Be careful about transporting compressed peat moss bales from the garden center. (They can make an awful mess if punctured accidentally inside a car!) And don't simply sprinkle powdery peat moss around. It should be completely wetted with the garden hose, then mixed well into existing soil. A common beginners' mistake is to pour dry peat moss into a dug hole, into a planter, or around plant roots then simply throw soil over the top and water briefly. After this procedure, the moss may stay in a compressed, dry heap below the soil for months (or even years) and prevent water from reaching plants. It needs a good soak to swell up and perform its aerating function.

The drawback to peat moss is that it's harvested from old bogs. Supplies are dwindling around the world and opinions differ about whether this is a renewable resource or not. Peat moss can also break down quickly in gardens, meaning that it's necessary to keep adding more. If you have concerns about the environmental impact of peat moss, use another soil additive.

Don't confuse sphagnum peat moss with dark, black peat (sometimes labeled "black muck" on bags). The black variety has decomposed further than its brown, dry cousin and looks terrific, but it has very little nutrient value and doesn't improve soil structure any more than regular peat moss.

Compost

There's a bewildering choice of commercially made composts on the market now. Bags at garden centers may be labeled "mushroom compost," "composted bark," "composted leaf mold," "composted lawn clippings," "composted woody materials," "composted manure" (from sheep, cattle, pigs, horses, chickens, or even crickets), "composted biosolids" (derived from human waste), and plain old "garden compost" (made from heaven knows what).

Which to buy? The best compost is usually the homemade kind because you control what goes into the bin, and usually that's a mix of organic materials. However, few gardeners ever manage to produce enough compost to spread all over their gardens. When buying a commercial product, bear these points in mind:

- Composts vary as much as topsoils. Try to find out three things: What is the compost made from? What are the levels of nutrients in it? And is the organic content high or low? Look for this information on labels.

- Generally speaking, composts made from woody materials or leaves are high in organic matter, but have low levels of nitrogen.

- Mushroom compost is high in nutrients, but often very salty, so should be used with caution.

- Composted animal manure comes in two kinds: *dehydrated* (or *dried*) and *fresh milled*. The former is the best kind to buy for gardens. It's more expensive than fresh manure, but has been dried at a high heat, killing weed seeds. While animal waste delivers high levels of nutrients, one drawback is that the manure is mixed with bedding litter, such as hay, which is frequently full of weed seeds. Dehydrated manure also has very little odor, while a whiff from a bag of fresh manure can knock your socks off. Bags of the dried kind are also lighter to lift into a car trunk.

- Cattle manure generally delivers less nutrients than sheep, horse, rabbit, or chicken manure, but many gardeners find it is absorbed more easily and is thus less likely to burn plants.

- Compost made from biosolids (human sewage) is high in nutrients but also highly controversial in some jurisdictions. Before buying it, check for proof that it has been treated properly and does not contain heavy metals or harmful pathogens.

The best kind of compost is homemade. Commercial composts vary and may emphasize one nutrient over another.

- No two commercial composts contain the same ingredients, even though they may carry identical labels. It pays to buy compost from several different sources and mix them together as sticking to one company's product may cause plants to be deprived of certain nutrients.

- If a bag of compost is very heavy, be suspicious. It has a high moisture content, a sign that it was poorly made. It also may contain sand or gravel.

- Homemade compost is far and away the best kind to use around plants. Irrespective of their type or size or growing requirements, all plants seem to do better with a bit of our own "black gold" added regularly around their bases. It's worth the effort to make compost. Many municipalities now make composting bins available at reduced or no costs. They also provide comprehensive instructions on how to compost properly. Be sure to check out these sources.

Coir

This product is fairly new on the market (sold under brand names like Soil Sponge or Peat Exceed). It's made of ground-up coconut fiber that's been dried and packaged in plastic. When mixed with water, a slab of coir expands to several times its size and thus makes an excellent substitute for peat moss. Environmentalists are singing the praises of coir because, unlike peat moss, it's a renewable resource. (Most coir sold in North America comes from countries such as Sri Lanka and Vietnam, which have huge stockpiles of the stuff.) Use coir only to lighten soil. It has virtually no nutrient value.

Turning compost several times a week will help the ingredients decompose. Open-sided bins make the job easier.

Sand

You'll often hear the term "sharp sand" mentioned in reference to gardening. This rather mystifying term means "coarse" or "builder's sand," which contains a lot of gritty particles and is thus a good additive for heavy soil because it breaks up the clumps and allows more oxygen to reach plant roots. Do not use sandbox or play sand or dig up sand at a beach or lake and bring it to your garden. These kinds of sands are too fine and may actually do more harm than good because they retain water.

Perlite and Vermiculite

These are usually mixed with peat moss and other materials and sold in bags marked "growing mix " But they're also available unmixed with anything else. Perlite, a volcanic material that's been heated to a high temperature, looks like nubbly bits of Styrofoam. Vermiculite, which is made from mica, is shinier and flakier. Neither has any real nutrient value, but they will lighten heavy soil and help retain water. Because they are dusty (and tend to be expensive), they should be used in conjunction with something else.

WHAT FERTILIZERS CAN AND CAN'T DO

It's easy to get hopelessly confused by fertilizers. Figuring out exactly what's in those big shiny bags and white plastic bottles sold at garden centers—and whether or not they're worth buying—can be a difficult task even for experienced gardeners. And while these additives can certainly make a difference to the way plants grow, it's important to understand a few things about them.

First, no amount of fertilizer—of any kind—can make up for underlying problems. If your soil is poor, or soil pH is out of whack, or if you've purchased weak or damaged or diseased plants, or if you've planted things improperly

Always read the fine print on fertilizer bags and containers. Follow instructions exactly or you may harm plants.

Understanding fertilizers

NUTRIENT	WHAT IT DOES
Nitrogen	Promotes leafy growth and stems and makes plants grow taller. Too much nitrogen delays flowering and fruit production, and results in excessive leaves and soft stems.
Phosphorus	Helps plants develop strong roots, flowers, and fruit.
Potassium	Aids growth of strong stems, disease resistance, and makes plants winter hardy. Helps fruit develop and ripen

and or/neglected them, simply tossing a dollop of fertilizer on top or into the watering can will not fix things. Nor will fertilizers correct climatic problems. If it's been too dry or cold during the spring, or if sunshine levels are unusually low or the climate is wrong for the type of plant you want to grow, applying fertilizers won't make a scrap of difference. In fact, they may do more harm than good because they place unnatural stress on plants.

Second, while fertilizers are often referred to as "plant food," they are only a small part of the plant-nourishing process. Green plants create their own "food" naturally via photosynthesis (sunlight, water, and carbon dioxide trigger them to release carbohydrates) and no less than sixteen elements or nutrients are required to perform this complex balancing act. The first three are carbon, hydrogen, and oxygen, derived from water and air. But the thirteen other nutrients are present in the soil—or at least they should be. If they've been leached out or need to be replenished or added, that's where fertilizers come in.

SIGNS OF DEFICIENCY	FERTILIZERS USED
Leaves are pale yellow and growth is stunted. Fruits may ripen prematurely.	Ammonium nitrate, ammonium sulfate, calcium nitrate, nitrate of soda, urea, fresh manure (especially from chickens, crickets, horses, sheep, and rabbits), fish emulsion.
Weak, stunted growth; leaves and stems may become purple; seedlings develop poorly and fruit ripens slowly.	Ammonium phosphate, rock phosphate, superphosphate, bulb booster, bonemeal, especially raw bonemeal.
Growth is slow. Plants are highly susceptible to disease and may develop "bronzed" leaves.	Potassium chloride, sulfate of potash (not recommended for young plants), greensand

Understanding Numbers on Fertilizer Bags

The most important basic nutrients that plants need are nitrogen (chemical symbol N), phosphorus (chemical symbol P), and potassium (chemical symbol K).

You'll see these three nutrients, often called macronutrients, represented as three numbers, in dozens of different formulations on fertilizer bags and other containers. They are always printed in the same sequence—that is, nitrogen first, phosphorus second, and potassium last. For example, 5–10–5 means the product contains 5 percent nitrogen, 10 percent phosphorus, and 5 percent potassium (or potash).

All plants need these three nutrients in a particular balance, which, of course, is different for every plant.

The Importance of Secondary Nutrients

These are calcium, magnesium, and sulfur, and in normal situations they occur naturally in the soil.

Calcium helps in the cell-manufacturing process and in early root growth.

Magnesium is required to develop seeds and chlorophyll, which enables plants to produce their green leaves.

Sulfur is a primary element in plant proteins and also helps plants develop a natural green color.

Other nutrients (often called micronutrients) need to be present in soil too. These include iron, manganese, copper, boron, zinc, and molybdenum.

When soils are deficient in any of the above, applying fertilizers can help—much in the same way that humans take vitamin supplements. But it is important to remember that, as with vitamins, they should be used sparingly.

What Fertilizers Contain

Products stocked by garden centers contain mostly significant amounts of the three primary nutrients—that is, NPK. When all three are present, they're known as "complete fertilizers." These may be sold in dozens of different formulations and packaged in a variety of ways as soluble powders, liquids, pellets, time-release capsules, spikes, granules, and foliar sprays. They're also marketed under names that are designed to entice us into buying a lot of different products: houseplant food, vegetable food, African violet food, bulb booster, shrub fertilizer, tree fertilizer, rhododendron food, lawn food. The list goes on. What's important to remember is that most of these products, irrespective of their names, contain the same basic three nutrients and that it often isn't necessary to buy a specifically named product for a specific plant. There's a lot of duplication and you can wind up buying three times what you need when the same product will suffice for several different types of plants.

The ratio of NPK that a fertilizer contains is always shown on the package or bottle (it's that list of three numbers, such as 5–10–15) because manufacturers are required by law to include it on their packaging. Check this list. You'll be surprised at how little NPK some fertilizers—often the fancy, expensive ones—contain. The higher the numbers on the bag, the bigger the bang for your buck.

Other fertilizers are designed to give a boost of secondary nutrients, such as calcium and boron. Many products also have herbicides included in the mix for weed control. Always read the information printed on labels before buying, and be sure that you are applying the appropriate product on your plants.

WHICH IS BEST: ORGANIC OR CHEMICAL?

This is where most of the confusion arises. Organic fertilizers are derived from natural, once-living sources, such as blood and bonemeal, compost, leaf manure, composted manure, fish, and seaweed. These products may be labeled in the garden center as "organic," "natural organic," "bio-organic," or "naturally derived." Whatever the label, the nitrogen in these products is the water-insoluble, slow-release type. This means that it lasts longer in the soil, is slow to have a beneficial effect, and usually will not burn plants.

Chemical or synthetic fertilizers, on the other hand, are manufactured from nonliving sources and are called "inorganic," "chemical," or "petrochemical" fertilizers. The nitrogen in these products is almost always 100 percent water soluble, so it's quicker to act, and plants feel the benefits faster. But it also may provide too much of a jolt, it washes out of the soil quickly, and its positive aspects aren't long lasting.

Confusingly, some common fertilizer products may be derived from both natural and chemical sources and yet are still labeled "organic." This is permitted in certain jurisdictions when a percentage of the product, but not all of it, is derived from water-insoluble materials. As a result, fertilizers marked "organic" or "rich in organic matter" may not be as "natural" as they claim to be.

> ## ALERT!
> *Fertilizers can kill*
>
> Always read instructions on fertilizer labels and follow them exactly. More plants are killed by amateur gardeners administering too much fertilizer than by virtually anything else. Don't presume that if you are more generous, you'll deliver twice the benefit. The reverse is more likely to happen; the plant will be overwhelmed and die.

In our environmentally sensitive times, most of us are inclined to pounce on anything marked "organic" because we presume it's better than chemicals. But there are pros and cons to both types of fertilizers. Some factors to consider when buying them:

SOIL BASICS 53

Organic:
- These usually contain quite small amounts of nutrients, which aren't in a convenient "balanced" mix of NPK. (Typically, they'll be heavy on one nutrient, but not others, so you may have to add something else too.) Check the labels.
- Nutrients are released slowly, and won't damage plants or their roots if you put too much on by mistake. (The exceptions are raw manure, which can burn plants if it is too fresh, and wood ash, which makes soil too alkaline in excess amounts.) However, because they take time, nutrients may not work fast enough to correct an acute plant problem.
- They need warm temperatures to start working but can be applied at any time in the growing season.
- Some, especially compost and manure, help build up soil structure and microorganisms such as earthworms.
- Those derived from sources such as seaweed and fish emulsion tend to be much more expensive than synthetic fertilizers.
- They generally do not harm the environment in their manufacture or use.

Chemical:
- They need only water to work and give quick results.
- They are usually convenient to use and mixed in precise proportions for plants' particular needs.
- They must be applied at specific times in the growing season and used with extreme care because nutrients are usually delivered in the form of salts, which can scorch roots and leaves and kill some plants.
- They are often cheaper than organic fertilizers.
- Over time, if used indiscriminately, they may impoverish the soil and create dependency on more fertilizers.
- They require vast amounts of nonrenewable energy sources to manufacture.

Most gardeners agree that it's best to opt for a balanced approach where fertilizers are concerned. That is, they regularly work natural fertilizers, such as manure and compost, into the soil around the base of their plants. However, they also add manufactured synthetic fertilizers occasionally to specific types of plants.

SOME PLANTS THAT BENEFIT FROM CHEMICAL FERTILIZERS
- *Roses:* Sold in liquid concentrates, pellets, and powders, most "rose food" products typically deliver NPK with an extra whack of phosphorus, which promotes bloom set on roses. Use them in spring.

- *Acid-loving plants, such as rhododendrons, hollies, azaleas, and some evergreens:* Fertilizers that contain chelated iron, aluminum sulfate, and soil acidifiers will promote growth and good color in these plants.

- *Spring bulbs, such as narcissi, tulips, and alliums:* A complete fertilizer that contains more phosphorus than nitrogen or potassium (typically in a 4–12–8 ratio) will encourage healthy root development and blooming. One product is called Bulb Booster. (Bonemeal, which has a similar phosphorus rating, is not as effective and is usually more expensive.)

- *Houseplants:* Fertilizers are sold in easy-to-use concentrated form, in liquids or crystals, so that they can be mixed into watering cans. They are formulated to provide a quick dollop of nutrients to potted plants and should be used every two or four weeks.

- *Container-grown plants, located outdoors:* These have similar requirements to houseplants because they are forced to grow in confined spaces, where soil nutrients are quickly depleted. A good all-purpose fertilizer for such plants is 20–20–20, mixed into a watering can every two weeks.

TIPS ABOUT SOME POPULAR PLANT CARE PRODUCTS

- *Antidessicants:* In northern climates, these sprays (or liquids) are useful for coating rhododendrons and evergreen shrubs to stop them drying out over the winter. Apply before freeze-up. Made of natural pine emulsion and nontoxic (although some products may contain chemicals too), they dry to a clear film. One application lasts for about three months. However, don't keep spraying dessicants on plants throughout the winter as they may permanently block their pores.

- *Corn gluten:* This is a hot new product that's gaining interest because it kills weeds, is 100 percent natural, and won't harm the environment. A by-product of cornstarch, it's usually sold as a fine yellow powder or in pellets in 5-lb or 25-lb (2-kg or 11-kg) bags. Sometimes corn gluten may also be mixed with soil additives such as bonemeal and potassium sulfate. Its primary use is on lawns, but it can be applied to stop weeds elsewhere in the garden. Because it's new, corn gluten can be hard to find at garden centers. Look in the weed and feed section. Some brand names are Bioweed, Corn Weed Blocker, Supressa, Safe 'n Simple, and TurfMaize. (See page 90 for instructions on its use.)

Organic fertilizers such as manure tea give a nutrient boost to vegetable seedlings.

- *Epsom salts (magnesium sulfate):* This is often recommended now as a boost for tomato plants and roses, sprinkled around their bases every two weeks. Dolomitic limestone is another good source of magnesium that's cheaper than epsom salts, and it needs to be applied only occasionally in spring. (Do not use it in conjunction with manure, as harmful gases are released.)

- *Leaf polish:* Most are not advisable because they are made with mineral oil, although some organic ones contain neem oil, which has a fertilizing effect. Their drawback is that they may leave a residue behind on foliage, clogging pores. Don't use milk, olive oil, or any other homemade potions either. A. shiny-leafed plant isn't necessarily any healthier than a matt one.

- *Mycorrhizal fungi:* This is a new environmentally friendly growth supplement that certainly helps perennials get established. Usually sold under the brand name Myke, it's a speckled gray, dusty substance, sold in plastic tubs, that's 100 percent organic. (See page 66 for instructions on its use.)

THE MAGIC OF MANURE TEA

Liquid feeding delivers nutrients to plants quicker than other methods. One safe, effective, environmentally friendly way to do this is to give them a drink of manure or compost tea.

Fill a burlap sack half full of animal manure. Use either fresh manure (in cities, zoos are often good sources) or a bag of composted manure from a garden center. Sheep manure is best, but other kinds are fine.

Tie a rope round the sack and suspend it from a bar stretched across the top of a big garbage bin filled with water. The sack should be completely immersed in the water. Leave for two weeks. Jiggle the rope now and then, so that the sack moves about a bit. If the manure is fresh (and a bit smelly), rest a lid over the top of the bin. If it's composted, you can leave it open.

The water will turn brown as dissolved nutrients seep out of the sack. Dilute this liquid with water until it turns a pale tea color. Pour around plants. It delivers a big dollop of nitrogen and is particularly useful around vegetable seedlings in spring.

An old burlap sack filled with manure (or compost) can be easily steeped in water to make manure tea.

Planting

4

Most of us enjoy planting, particularly if it's a warm spring day and we've just spent a fortune on plants at a garden center! However, in our zeal to get things into the ground, we often pick the wrong spot and rush the task, so our purchases don't get off to a good start.

That's too bad because plants will experience less problems (and wind up being vastly less work) if they are properly positioned and planted. Bugs and diseases are far more likely to attack plants that are forced to grow in an environment that doesn't suit them.

WHERE TO PLANT

Deciding on the right place for the right plant can be surprisingly difficult, which is why so many gardeners wind up afflicted by a malady called TUTMEA—the urge to move everything around. Remember that plants spread, and some will grow very tall, often remarkably quickly. Very few plants keep the diminutive dimensions that they have when they're sitting in plastic pots at garden centers.

It's easy to make the mistake of coming home, taking newly purchased plants out of their containers, then eagerly plonking them too closely together in the flower bed. Many of us are also guilty of buying a new plant on impulse and squeezing it into a space that we secretly know will wind up being too small. The end result of such hasty handiwork is that plants don't get a fair chance to develop. The bossy ones crowd out the weaker ones and none of them

Left: Don't rush planting. Take the time to pick the right site and dig a hole that's deep and wide enough. Plants will be healthier and less work if they are planted properly.

receives sufficient light or air. (Inadequate ventilation around plants is a very important factor in promoting the spread of fungal disease.)

Before planting anything, the watchword is *Whoa*. Make a cup of coffee, sit down, and take a hard look at the space where you intend to plant your purchases. (One well-known garden writer calls this "creative staring," and she does a lot of it.) Figure out where everything will get a chance to grow well. Better still, decide what plants you need *before* going to the garden center. And don't get too many. (See page 29.) Take the time to read all the tags stuck into containers of plants you buy. The information provided on these tags is improving all the time. Most now give details on how wide and tall plants are likely to be when they reach maturity and what their light requirements are. Don't ignore the information on the tags. Although a plant's final size depends upon lots of factors (such as climatic conditions, position, soil, and fertilizer) and growers' predictions often fall short of the mark, if the tag says "Mature height: 36 in (91.5 cm); Spread: 18 in (46 cm)," you'd better believe it may happen and plan accordingly. A carpenter's metal tape measure is useful when allocating space for plants. So is drawing some diagrams before planting.

More Room, Not Less

When you aren't sure how much space to allocate a certain plant, it's better to err on the cautious side and give it more room than you think it will need, instead of squishing it into a tight spot. This will avoid the necessity of moving or dividing that plant and/or its neighbors (often a tiresome chore, which disturbs the plants) a couple of seasons later. If you're establishing a flower bed around a new home, and are putting in perennials for the first time, don't be discour-

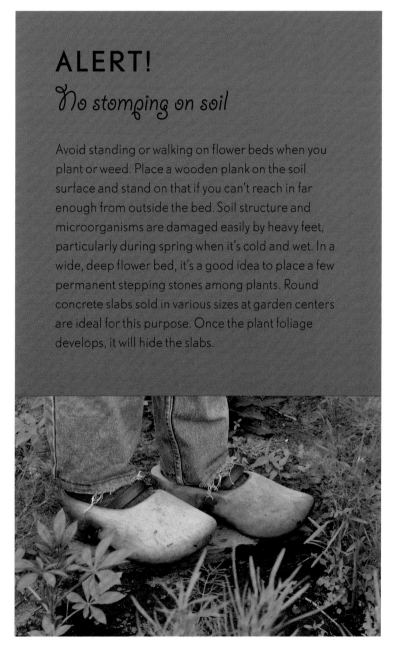

ALERT!
No stomping on soil

Avoid standing or walking on flower beds when you plant or weed. Place a wooden plank on the soil surface and stand on that if you can't reach in far enough from outside the bed. Soil structure and microorganisms are damaged easily by heavy feet, particularly during spring when it's cold and wet. In a wide, deep flower bed, it's a good idea to place a few permanent stepping stones among plants. Round concrete slabs sold in various sizes at garden centers are ideal for this purpose. Once the plant foliage develops, it will hide the slabs.

aged if you follow instructions to the letter and then everything looks a bit skimpy the first season. Most newly planted flower beds do, but those bald spaces will fill out within a couple of years. In the meantime, gaps between clumps can be covered with decorative mulch, some annuals, or a temporary container of cheery annuals like petunias or geraniums.

WHEN TO DO IT

Spring or fall? Or any time? Much depends upon the location, the type of plant, and how hot the weather gets.

In frost-free regions of the South, planting during fall is best because in spring plants don't have enough time to get established before getting zapped by the torrid heat of summer.

In northern climates, however, some gardeners prefer spring. Others think fall is preferable. Many insist that you can plant at any time in spring, summer, or fall, as long as plants are kept mulched and well watered. Some points to remember:

- Because perennials stay outside all year, they can be planted and moved early in the spring as soon as the ground has thawed. In northern regions, this may be any time between the beginning of March and mid-May. However, avoid digging in soil that's still very wet, cold, and clumps together a lot, which often happens in clay soils. Wait till it's warmed up a bit.

- Many perennials also take well to fall planting as long as it's at least a month before heavy frosts set in. In some northern areas, this means you may be planting in late August. (An exception is areas of heavy clay soil. See page 42.)

- Don't plant perennials during heat waves or if there's been a recent long, hot, dry spell.

- Pick a cloudy day for planting or, better still, plant in the evening. Never subject anything that's newly planted to an immediate dose of hot sunshine.

- Time planting for when rain is forecast. Most plants seem to settle in better with a drink administered by Mother Nature rather than the garden hose. (And it's a wonderful feeling to snuggle up in bed at night after a day of planting and listen to the rain giving everything a good soak.)

Contrary to popular opinion, it's also safe to plant when it's actually raining, provided the plants aren't handled too much (avoid breaking leaves and stems because it invites disease and insect problems) and you don't work wet soil too much or keep walking on it.

- Wait till frosts are completely over and the soil has warmed before planting annuals such as impatiens, petunias, coleus, cosmos, celosia, *Tagetes*, and Nicotiana. The same goes for most vegetable and annual herb plants. Tomato plants, basil, cucumbers, and all types of squash seedlings are extremely sensitive to frost. Potato pieces won't develop into healthy plants if they're buried in cold, damp soil.

- Cool-weather annuals such as pansies, snapdragons, and old-fashioned marigolds (Calendula) can tolerate a couple of degrees of frost with no harmful effects. You can also put established geranium plants (pelargoniums) out early in spring with no ill effects.

IF JACK FROST MAKES A RETURN VISIT

With North America's increasingly topsy-turvy weather patterns, we can no longer depend upon the air temperature to gradually progress from cool to mild to hot after the winter. In many areas now, unseasonably hot weather may hit in spring, then it gets abruptly cold again with a few nights of surprisingly heavy frosts, often into June.

If the temperature nosedives and you've already planted, don't panic. Here are some tips on saving frost-sensitive specimens:

- A dry, light cotton sheet (or blanket) thrown over a flower or vegetable bed at night is sufficient protection from light frosts for most plants. Every plant must be covered by the sheet (stems and foliage sticking out will get zapped). Heavy cloth tends to crush plants.

- Do not use sheets that have been rained on and are still wet. They will freeze and have the effect of encasing the plants below in an icebox.

Upturned plastic flower pots and cardboard boxes protect new plantings from late spring frosts.

- Buckets of water placed at intervals throughout the garden may stop light frosts from inflicting damage.

- Overturned plastic pots, empty plastic bottles with their bases removed, and cardboard boxes all make good plant protectors.

- Do not use plastic sheeting or garbage bags. They don't stop frosty temperatures from getting through.

- Remove protective coverings in the morning once the air temperature has started to rise. Do not leave pots and boxes sitting over plants all day, particularly when it's sunny.

- Do not cover spring bulbs, such as tulips and daffodils. If frost hits, their flowers will simply close up and go semidormant until the weather improves. Throwing blankets over the top does more harm than good because it breaks and bruises stems, leaves, and flowers.

- If plants' leaves start curling up at the edges or leaves and stems turn black or droopy after a cold night, they've been hit by frost. Newly planted perennials may look pinched, but they'll recover. (You may have to snip off frost-zapped foliage and stems.) However, most annuals won't recover, and it's not worth trying to cosset them back to life. Wait a few days, inspect their progress, and if they're still drooping, wizened, or turning black, throw them out and buy new ones.

- When the first fall frosts arrive, you can prolong the growing season for annual flowers and vegetables by throwing sheets over them for the night. This works particularly well for ripening tomatoes.

Top: Do not cover tulips that have been zapped by frost. They will recover on their own.
Bottom: Extend the growing season by throwing cotton sheets over vegetables on frosty fall nights.

ALERT!
Watch those snow shovels

In northern climates, avoid planting anything too close to pathways where you'll be shoveling snow. The branches of shrubs such as burning bush *Euonymus alatus*, *Weigelas*, and *Spirea* can be badly mauled by misplaced whacks with a shovel during winter. So can the thin, delicate stems of clematis vines. Perennial plants with shallow root systems, such as iris, may also be inadvertently yanked right out of the ground, and you usually won't notice the damage till spring when the snow melts.

Road salt used on pathways and sidewalks to melt ice also wrecks many plants. Agricultural urea (sold in pellet form at many garden centers) is an expensive substitute that will melt ice as effectively as salt. Urea has a beneficial effect, too, as nitrogen will leach into the soil next to the pathway. Buy the slow-release kind to avoid burning plants.

HOW TO PLANT MOST THINGS

- Dig a hole that's several inches wider and deeper than the container that the plant came in.

- Unless you have perfect soil (and few gardeners do), add a scoop of homemade compost or some bagged composted manure. Decomposed, chopped-up leaves from the garden are also good. How much depends upon the size of the plant. Whatever you use, don't overpack the hole, so the plant winds up sitting too high above the surface of the soil.

- Fill the hole with water from the garden hose. Let it all soak away before planting. This will settle the compost. (If draining away takes ages, your soil is too clayey and it could benefit from soil amendments. See page 44.)

- Turn the pot over and remove the plant *gently*. Don't yank on the stems to get it out. In some cases, it may be necessary to thump on the bottom of the pot, prod the root ball with an instrument stuck through the hole underneath, or run a sharp knife around the pot's sides. With cell paks, simply push up the bottom of the pak and the plant will pop out.

- If the plant comes in a fiber pot, *don't* simply slash the sides of the pot and stick the whole thing in the ground. (Much misleading information is disseminated on this issue.) If the pot is made of thick fiber, it may not break down in the soil for years and in the meantime, your new plant's roots will be cramped and unable to grow properly. (Plants should be put into the ground, fiber pots and all, only if they are large and too complicated to maneuver when taken out of containers. This may happen with shrubs and trees, but not, in most instances, with perennial plants.)

- Separate tightly packed roots with your fingers or a small tool. Trim off any mushy or blackened bits. With annuals, it's perfectly okay to pull the roots apart or cut right into them with a knife. This won't affect the plant's ability to grow. In fact, such "savage surgery" often makes annuals grow faster.

- Position the plant in the hole so that it's *at the same level* as it was in the pot. Make sure roots aren't all bunched on one side (distribute them evenly) and that the stem is vertical, not tipping sideways. With bare-root plants, spread roots out in a circle.

- Fill in with soil and tamp it down. If the plant is too high or too low, dig everything out and start again. (Yes, this can be a chore, but it's important to get the planting level right.) With large perennials, shrubs, roses and trees, fill in the hole only halfway, then water deeply. After that has drained away, add the rest of the soil.

- Make a moat by piling up a little ridge of soil in a circle around the plant with your hands or a trowel. Be sure water doesn't dribble out the sides of this moat when you give plants a drink.

- Put some mulch around the base and water thoroughly. If you find that the soil level sinks after watering, you may need to top it up. Keep soil moist for at least a couple of weeks, but don't overwater. Push your finger into the soil occasionally, to a depth of about 1 in (2.5 cm). If it's dry down there, water. If it's still damp, don't. (With shrubs and trees, give them a good soak once a week with the garden hose.)

IMPORTANT PLANTING TIPS

- If you're putting in a new flower bed, arrange plants, still in their pots, on the top of the bed before digging any holes. This can be helpful in deciding where everything should go. Remember that tall plants go at the back, so they don't stop the sun from reaching smaller ones.

- A trowel with a metal blade is the best tool for planting most annuals and perennials. Garden spades tend to make holes that are too big, and are better for shrubs and trees.

- If tree roots are extremely dense (a common problem with maples), you may have to forgo planting in the ground and settle for containers or planters placed around the base of the tree instead. (See page 68.)

- Applying a mild fertilizer to annual plants while they're still in their cell paks or pots can get them off to a good start. Mix a product like 20–20–20 into a watering can at *half the strength indicated on the package or bottle* and drizzle this on the pots a day or two before planting.

- Blood meal, bonemeal, and fish emulsion are often touted as good organic fertilizers, but don't add them to planting holes if your garden is visited by squirrels, cats, dogs, or other scroungers. Modern bonemeal has very little nutrient value anyway. (See page 156 for more on animal damage.)

> ## ALERT!
> ### Tags make good memory joggers
>
> It's easy to forget what plants you bought and where you put them. Sticking tags into the ground beside new plantings will help you remember—initially, anyway. However, avoid festooning the garden with too many permanent plant tags because it's not very attractive.

- An application of mycorrhizal fungi (see page 56) will help perennials get established. Pour a small scoop of it into planting holes. It must make contact with plant roots to have any beneficial effect (do not add it to the topsoil after planting) and works best on plants that will be growing in poor soil. Because it is expensive, Myke is not worth using with annuals.

- Slow-release granular fertilizer can also be added in small amounts to planting holes. These particularly benefit annuals, which are fast growing and need lots of nutrients. (See page 49 for more on fertilizers.) There are also some kinds of new slow-release fertilizers that can simply be sprinkled on top of the soil. But don't overdo these, because in hot weather, they will form a crust and not disintegrate properly.

- Gentle watering is necessary for new transplanted annuals and perennials, especially in the early days. Use a hose with a watering wand. Don't train a strong jet of cold water on them.

- Small plants and seedlings go through less transplant shock if watered with room-temperature water that's been sitting around for a few hours. Use the bucket and yogurt container method (see page 84).

- While mulching new plantings is generally beneficial, don't do it when the weather's very wet because it encourages slugs and snails (see page 143–45).

- Opinions differ about the virtue of pinching flowers off newly planted annuals. Some experts say that removing these early flowers (leaving only buds behind) promotes bushier growth, stronger root systems, and more flowers later on. Others say that's nonsense and that the annuals can go into the ground as is, with their existing flowers intact. The choice is yours.

- Annuals and perennials frequently get very leggy while sitting around waiting for buyers in garden centers. Generally speaking, you can cut most of them back by half after planting, with no ill effects.

Newly planted annuals prefer room temperature water (or manure tea as shown here.) Fill a bucket and dip a small container into it.

PLANTING UNDER TREES

Trees are wonderful to have in gardens because they add character, structure, and coolness in summer, but they also create the awkward phenomenon called "dry shade." Homeowners with small city lots usually become familiar with this problem the moment they try to grow something beneath their trees. The new plants fail to flourish because of three basic obstacles: the trees shut out sunlight, they compete for moisture with the plants, and the tree roots hog the soil, depleting its nutrients. Roots in fact take up a surprising amount of space near the soil surface. Contrary to what we're taught in school, most don't plunge deep into the ground. They ramble horizontally, within the first couple of feet of soil and the poorer the soil, the closer roots are to the surface. In many gardens, they're often in a huge tangle, making it difficult to find room for anything else.

Before planting under trees, try to work in as much organic matter as you can—the more, the better. Good soil preparation is particularly important under maples and conifers (the latter's acidic needles are hostile to the development of many plants). If trees are large and established, snip some of the roots out of the way with garden pruners, in order to work in the soil. (This won't harm the tree, although it's not advisable with recently planted saplings.) Keep the soil moist by mulching it and replenish the soil every spring by adding another heavy layer.

Under deciduous trees, it's preferable to plant in early fall because the canopy of leaves is thinning out at that time and plants have a better chance of getting established when some sunshine can reach them. Water well, and don't stop regular watering for several weeks.

If tree roots are extremely dense (a common problem with maples), you may have to forgo planting in the ground and settle for containers or planters placed around the base of the tree instead. A decorative mulch of composted tree bark or gravel also looks attractive.

See page 22 for recommendations on plants that will grow under trees.

THE PROBLEM OF WIND

Wind isn't all bad. It ventilates plants, keeping them healthy. It dries out sodden soil and disperses pollen, seeds, and pleasant scents. And there's nothing like a good breeze to refresh everything and everybody. Without some wind, a garden would seem lifeless and stagnant, and be more prone to bugs and diseases.

However, the problem with wind is that often it's too fierce. Increasingly, gales and "twisters" erupt suddenly in many areas of North America, and on the Prairies strong winds make their presence felt most of the year. If you live in an exposed location, be prepared to keep replenishing your topsoil (because it will blow away steadily) and to install windbreaks, such as fencing and hedges. Plants are particularly vulnerable to wind damage right after they've been planted. Pick low-growing types and go for lots of evergreens. (Some toughies to try: Colorado blue spruce *Picea pungens* var. *glauca*, a low-growing juniper called 'Arkadia'; Swiss stone pine *Pinus cembra*, Mugo pines, hedge cedars, and Siberian cypress *Microbia decussata*.) Do not grow the popular vine clematis in areas that are buffeted constantly by winds, as most varieties have very thin, fragile stems. When cultivated in containers on the balconies of high-rises, clematis stems can be completely shredded by the wind.

DIFFERENT BULBS NEED DIFFERENT TREATMENT

Spring-flowering bulbs should be planted in fall. Narcissus, alliums, crocus, fritillaria, and scillas prefer to go into the ground in September and October, but you can procrastinate with tulips. As long as the ground can still be worked, tulips won't object (although don't forget that fingers get frozen when scrabbling in cold soil in December).

If the fall is unseasonably warm, wait to plant all spring bulbs until *after* a light frost. Tulips are particularly sensitive to temperature: they need a cool period in the ground immediately after planting and may not send up flower stalks the following spring if they go in too early.

In southern climes, in order for these kinds of bulbs to perform successfully, it's necessary to subject them to a cool period of three to four months in a refrigerator before putting them into the garden. Do not store fruit in the same crisper drawer, as some fruits (and tomatoes)

ALERT!
Protecting plants from wind

Small plants, particularly annual seedlings, need lots of TLC right after planting. Even if it's not windy, protecting them with recycled plastic or fiber pots is smart. Cut out the bottoms of pots with sharp shears or scissors (1-qt/1-L, 1-gal/3.75-L, and 2-gal/7.5 L pots all work, depending upon the size of the plant), then simply place these bottomless pots over the top of the plants, pushing their edges firmly into the soil all the way around the plant.

These makeshift plant protectors can be left in place for several weeks. When plants are established, slip them gently off, being careful not to damage developing foliage and stems.

Bottomless pots are also an effective way to deter nosy cats and squirrels from digging around an area that's newly planted. (See page 156 for more on animal damage.)

give off ethylene gas, which may kill the bulbs or stop them from flowering. Plant outside in January. They will flower for only one season and then die off.

Summer-flowering bulbs originate in the tropics. That means they must be planted after your region's last frost-free date. However, because many summer bulbs (also corms and tubers) take weeks to develop and come into flower, they are best started indoors in northern climates, or bought as started plants from garden centers.

Begonias, caladiums, calla lilies, cannas, and dahlias will all perform faster if given a head start indoors (by you or the garden center's growers), then placed outside after frosts have finished. Gladioli, however, should be planted directly into the garden once the temperature gets above 55°F (13°C) and, in most northern regions, they won't send up their tall flower stalks until late summer. (In southern climates, gladioli can be treated as perennials and stay in the garden year round.) *Gladiolus callianthus* (also known as *Acidanthera*), which produces smaller flowers than its garden-variety cousins, is a good choice of gladioli for small northern gardens and for balcony gardeners. If planted in May, it will produce gorgeous white blooms on long, elegant stems by the beginning of August. It doesn't require full sun and makes a stylish container plant.

Eucomis, commonly known as pineapple lily, is another summer bulb that doesn't need months to come into bloom. Reliable and easy to grow (provided it gets a good burst of heat to get things going), it quickly sends up a stalk enveloped in little flowers and topped by a crown of leaves, which somewhat resembles a pineapple. *Eucomis* makes a good container plant on a deck too. It will get by nicely without full sun, but if the spring is cold, it will sulk.

Because they must be dug up in the fall and brought indoors for the winter, many summer-flowering bulbs are best treated as container plants rather than put directly into garden flower beds.

How to Plant Bulbs

- Dig deep holes. When planting a lot of bulbs in a group, it makes more sense to remove a trench of soil 9 in (23 cm) deep and 2 ft (.6 m) wide with a spade than to make individual holes with a trowel or bulb tool called a dibber. Depth is generally recommended to be three times the size of the bulb, but if you're bothered by squirrels or live in an area that experiences frequent freeze-and-thaw cycles during winter, dig deeper. (Remember that bulbs planted deep in the ground will bloom later the following spring. Don't panic if they take a long time to show.)

- Digging holes in clay soil is hard, but avoid the temptation to make shallow holes (as is recommended in some gardening books). In northern areas, during winter freeze-up, heavy clay soil may crack open and pop bulbs right out of the ground. They need to be planted deeply to survive.

- It can be difficult to squeeze bulbs into a flower bed that's already full of plants, shrubs, or tree roots. Use a pointed metal trowel or dibber to excavate holes, and snip off tree and shrub roots if they get in the way. If the trowel won't go more than a couple of inches deep, find somewhere else to plant the bulbs.

- Make sure the area is well drained. Bulbs hate getting waterlogged. In areas of heavy clay, add compost and a layer of sand, then place the bulbs on top, pointed end facing upwards.

- Although many experts advocate spacing bulbs several inches apart, most look better planted in tight groups.

- For a swath of bulbs in a lawn, don't dig holes in regimented rows. Clumps are better. For a naturalistic effect, some gardeners like to take a handful of bulbs, fling them on the ground, and then plant the bulbs where they fall. (But in a lawn that's densely packed with grass roots this can be an impossible task.)

- Add a bulb fertilizer, such as Bulb Booster, to holes or the planting trench. Modern bonemeal, although widely used, goes through so many manufacturing processes that it provides very little in the way of nutrients. (It also attracts animals.)

Triple decker does it!

In a small city garden, space can be at a premium. If that's your problem, remember that it's perfectly okay to plant spring bulbs in layers. For instance, excavate an area deeply and put a layer of late-flowering tulips on the bottom. Toss some soil over those, then add a layer of mid-season tulips. Add more soil and top things off with a layer of early-blooming narcissi or crocus, which are the first spring bulbs to pop up.

You can also, in a tight spot, plant spring bulbs *underneath* perennial plants like sedums and euphorbias. The bulb stems and leaves will find their way up through the perennials' roots with no ill effects. Then when the bulbs finish blooming, the plant foliage will take shape and mask the appearance of those dead bulb stems and leaves.

- Some people add human hair or prickly pieces of holly to the planting hole to deter squirrels. Another idea is to encase the bulbs in an empty coffee can with holes punched in the bottom and hardware cloth or wire netting placed over the top. Do not use mothballs (toxic to pets and wildlife) or cayenne pepper, which can blind squirrels and cause them incredible pain. (Humane societies now ask us not to do this.)

- Smooth the area out flat with your hands after planting (disturbed soil is more likely to attract squirrels and other curious critters) and water thoroughly.

- Bamboo barbecue skewers stuck close together into the soil with sharp ends facing upwards are one of the best ways to stop animals from digging there. Sharp gravel sprinkled on top or a layer of wire netting may also work, although wire netting is fiddly when you want to pull weeds or work the soil. If you live in an area where native North American buckthorn trees grow, cut some branches and place those over the top of the planted area. Very few animals are brave enough to venture in among buckthorn's vicious needle-length thorns.

PLANTING IN CONTAINERS

- Do not use garden soil or bags marked "topsoil" sold at garden centers. Buy potting soil or "container growing mix" (labeled as such on bags). Plants grown in containers need a lighter growing medium than in the garden and specially formulated mixes contain what they need: peat moss, vermiculite or perlite, possibly composted manure, and other ingredients such as bentonite clay and hydrogel granules to retain water.

- If you have lots to pot up, it can be cheaper to make your own mix. Combine a bag each of composted sheep or cattle manure (or some of your own compost), peat moss, and vermiculite or perlite in a garbage bin or wheelbarrow. Mix thoroughly. Some gardeners also add a handful of slow-release fertilizer (be sure to follow manufacturers' instructions as to quantity). A few scoops of builder's gritty sand are beneficial, too, for herbs, cactus, and bulbs.

- Cut bags open and feel the mix. If it's dry and flaky (those that contain lots of peat moss usually are), tip a large jug of water into the bag and let it sit overnight.

- Put a coffee filter over the holes in the bottom of containers (pieces of crockery or broken-up pots aren't necessary) and fill them only two-thirds full. Then arrange plants on top of the mix (still in their little plastic pots). Keep moving them around. Don't take plants out of pots until you decide on an arrangement that satisfies you.

- Use plenty of plants. They can be closer together than in a regular garden. Position tall plants in the center (or at the back of a window box planter). Trailing ones are best around the edge.

- When ready to plant, loosen root balls with your fingers, then position the plants in containers, making sure they're not too high. (See page 64.) Then tuck more soil in around plants. Leave 1 in./2.5 cm between the soil surface and the pot rim, so that when you water, it won't overflow and make a mess. (On balconies, it is a good idea to put catchment saucers beneath the pots.)

- Water deeply. It should come out of the drainage holes.

- Avoid putting containers in hot sunshine right away. Leave them in the shade for a day or two, then move. If containers are too big to haul around, shield the plants with a light cloth or folded newspapers.

- Fertilize regularly because nutrients get depleted quickly in containers. A good regime is 20–20–20 mixed into a watering can every two weeks (follow manufacturer's instructions).

Plants can be densely packed in containers. This one contains (left) Lamium maculatum, dragon's wing begonias, and coleus.

- Be aware that this kind of gardening, while hugely popular, isn't as low maintenance as many people claim. Containers dry out easily and will require constant watering in hot spells.

PLANTING ROSES IN COLD CLIMATES

Much misleading advice is disseminated about the depth at which to plant roses. The fact is, regions that are Zone 5 or colder, you should plant them in a *deep* hole.

Look for the bud union on the rose plant. This is the rounded lump that sits on top of the center stalk and roots. (It has branches sticking upwards from it.) Although plant labels often advise planting the bud union at soil level (or above it), in cold climates this should be buried at least 2 in (5 cm) below the ground, and in areas of extreme temperature, plant as deep as 4 in (10 cm). Failure to do this may condemn the rose to death over the following winter, or it may languish and never thrive.

In very cold regions, roses are best planted in spring. Make a wide, deep hole and enrich it with compost. Water several times after planting. Most roses require full sun, rich soil (add compost around their bases every spring), and excellent drainage. They don't like to dry out and prefer a long soak with the garden hose once a week rather than daily dribbles from the watering can.

Container-grown roses are the easiest to plant. Bare-root roses should have their roots soaked in water for a day before planting.

Rugosa roses are the toughest kinds to grow. They cope well with cold and salt water and native varieties thrive along the northeastern seaboard of North America. But many rugosas grow into large bushes and are thus not suitable for small gardens. (Check with the garden centre before buying.) The native kinds bloom only briefly in spring, but the quilted pattern on their leaves is interesting, they have a lovely fragrance, and their rosehips, produced in late summer, are beautiful—the color and size of cherry tomatoes. Small "fairy" roses, which seem to survive anything and are sold in garden centers everywhere, are good choices for containers. Do not plant hybrid tea roses if you are a lazy gardener who dislikes having to coddle plants because most roses require extra care.

See pages 110, 146 and 151 for rose diseases and insect problems.

One drawback to growing plants in containers is the need to water constantly.

THE MARVEL OF MULCH

Twenty-five years ago, most people had never heard of mulch. Now the practice of placing a protective layer of material around the base of plants, shrubs, and trees is standard procedure in gardening. Everyone mulches, and all types of mulch—from buckwheat hulls to weed-control mats made of synthetic fibers—are sold at garden centers. It's obvious that mulch mania is here to stay.

Mulch is certainly marvelous stuff. It smothers weeds. It protects plants from heat in summertime and from freeze-thaw cycles in winter. It softens up hard-as-a-brick soil in an almost magical way. It encourages earthworms to multiply. It's a boon when you've just planted something because it conserves moisture. And organic mulches may have some nutritional benefits. However, you can have too much mulch. There are also right—and wrong—ways to make use of it.

How to Use Mulch Properly

More mulch doesn't necessarily mean better. Most beginner gardeners, in their zeal to protect plants or banish weeds, overdo it. Drive around any subdivision and you can see mulch piled high around the trunks of newly planted trees. This is *not* a good idea for several reasons.

Generally speaking, layer no more than 4 in (10 cm) of organic mulch around the base of plants, shrubs, and trees. More may suffocate the roots. Thick layers can also encourage mold and other fungal diseases to develop. (Some botanical gardens that have depended heavily on mulch in recent years are now having problems with such diseases.) Organic mulch is preferable to other kinds because it will eventually break down into the soil and is thus kinder to the environment.

Do not let any kind of mulch touch plant stems or rosettes of developing leaves *at all*. Ideally, mulch should be placed at least 1 in (2.5 cm) away in a circle around the stems or trunk. (This can be difficult with a mulch of leaves in the fall because they're bulky and awkward to position properly. That's one reason it's best to chop up leaves into a finer material if you can.) Around trees and big shrubs, it is extremely important not to let the mulch touch the trunk. If it does, new tree roots will push out into this material instead of into the soil below.

If you're buying bagged mulches, about thirteen bags of 2 ft^3 (.05 m^3) each will cover an area of 100 ft^2 (9 m^2) to a depth of 3 in (7.5 cm).

Mulch types include the following.

Leaves

Leaves are the cheapest and easiest for city gardeners. Simply rake and sweep them up in fall. (If you have a huge garden, sneak out and swipe the bagged leaves that other homeowners put out for city composting trucks because you'll need lots.)

To get newly fallen leaves to break down faster, run a weed whacker or lawn mower (set the blade on a high setting) over a pile of them. You can also stuff leaves into dark-colored garbage bags, add a scoop of soil, tie the bags, and leave them outside in an unobtrusive spot for the winter. They'll break down amazingly quickly, then they can be used as mulch the following season.

There is no truth to the rumor that oak leaves make soil too acidic. Truly huge quantities are required to have that effect. However, oak leaves are as tough and durable as Styrofoam plates, and they can take years to decompose. Avoid them if other leaves are available.

Some fallen leaves, particularly those of maple trees, have a tendency to become matted and turn moldy, especially if they haven't been shredded first. Don't layer them too thickly on flower beds.

Cocoa Bean Fiber

Now a garden center staple, cocoa bean fiber adds nutrients to the soil but can be expensive, so it is best mixed with something else (like sheep manure or leaves). Avoid the lurid orange kinds, whose colors tend to clash with plants. (And contrary to popular opinion, these colors don't occur naturally; the fiber is dyed.) Buckwheat hulls, which are lighter, are sometimes available too.

Coir

Coir is fairly new and is sold under brand names like Soilsponge and Peat Exceed. Basically a pulverized version of cocoa bean hulls, it blows around too much to use as mulch on its own, but is fine mixed with something else.

Compost

This is terrific, and there are all kinds now (see page 46). The homemade variety is best, but you'll find you may not be able to make enough of it to use throughout the garden.

Pine Needles
Pine needles are okay, but they become very dry, so do not use them in areas where fires could be a hazard. They also tend to be too acidic for most plants.

Straw or Hay
Straw makes the best kind of mulch in a vegetable patch because it's usually weed free. A couple of small oblong bales is sufficient for most needs. Hay is less satisfactory than straw because it usually contains weed seeds. If you use it, make a layer at least 6 in (15 cm) thick. The exception is salt hay, which consists of grasses harvested from salt marshes. Seeds of this hay won't become a problem in most gardens because they require wet, salty soil to germinate.

Neither straw nor hay looks very attractive when used as mulch in a flower bed, however.

A thick layer of straw will stop weeds growing around vegetable seedlings.

Wood Bark/Chips/Nuggets
Shredded bark is fine, but do not use large, chunky chips or nuggets on flower beds. They tend to pack down and fuse together, stopping moisture from reaching plants.

Stone Mulch
Chips of granite, limestone, or marble—or river gravel—are now available in bags. They make an acceptable mulch around shrubs and some rock garden plants, but look too "industrial" in a flower bed.

Landscaping Fabric
This is usually a black or gray see-through material that is sold in rolls and now available at most garden centers. The fabric is great to spread over large areas to prevent weeds from developing or on slopes to stop erosion. It also works as mulch in a new flower bed—simply cut holes for plants and shrubs, but don't use this cloth if you plan to work the soil afterwards because you won't be able to get at it.

Old Asphalt Roofing Tiles and Carpeting
Use in areas where you aren't worried about appearances, such as a vegetable patch. Both make great pathways between rows of plants. But avoid placing

ALERT!
Don't wait too long to mulch

Be careful if you're using a pile of leaves from the garden (or some homemade compost) as mulch during the winter months. If you wait too late in the year to lay this stuff around your plants, it may have turned into a frozen lump and become impossible to move.

Laying down a sheet of black plastic will kill unwanted plants growing underneath it.

either tiles or carpeting too closely around growing vegetables as they stop water from reaching the roots. And don't recycle old foam underlay as mulch. It will crumble and make a mess in the garden.

Black Polyethylene Plastic

Sold in rolls at hardware stores, this is cheap and excellent for smothering large areas of weeds. Punch holes in it and plant vegetables. Many, particularly squash, will perform beautifully. But black plastic isn't aesthetically pleasing or practical for most flower gardens. It's also not very environmentally friendly. Look for the newer plastics, made of GMO-free corn, which are biodegradable.

Mulching in Summer

In summertime, mulch can be a real boon, keeping weeds down and soil moist. It will also help break up clay soil. And if it's an organic mulch of something like compost, some nutrients may leach down into the soil, benefiting plants. However, in prolonged wet weather, mulch may stay too damp. Besides promoting fungi, sodden mulch becomes a perfect hiding place for slugs and other pests. If you have a problem with creepy-crawlies, peel back the mulch and take a look. In serious infestations, remove it entirely and stick to growing plants in bare soil.

Mulching in Winter

In cold regions, mulching most borderline hardy plants (or recently planted ones) for the winter is a good idea. It doesn't stop plants from freezing solid (there's a lot of confusion about this), but it does soften the impact of freeze-thaw cycles. Violent temperature swings are increasingly hitting many areas of North America, and if soil is exposed during a mild spell, plants may start growing and then get zapped by the returning frost. Timing is important, however, when

laying down winter mulch. Wait until after the ground has frozen. Don't do it while the soil is still workable. In windy areas, weigh the stuff down with branches or bricks, and be sure not to smother plants with mulch.

Composted leaves make a great mulch around plants in fall.

When Spring Comes

Once the weather starts to warm up, don't leave winter mulches around plants. It's important to peel back these layers to give plant roots a chance to warm up, otherwise their growth spurt will be delayed or hindered (and old wet mulch may invite disease and insects). New shoots shouldn't have to struggle up through a layer of mulch to reach the light. This is particularly crucial for spring bulbs, which are often blanketed in thick layers of last year's leaves, and then send up spindly, weak stems as a result.

And if you want to restore a protective layer of mulch for the summer, let the plants get going properly first. They need to develop some fresh new growth before mulch is piled up around them. If the spring is very dry, water before mulching. It's also easier to add granule-type fertilizers around the plants before laying down mulch, but water-soluble kinds can be administered afterwards.

ALERT!
A free source of straw

Keep a sharp eye lookout at supermarkets and garden centers in late fall. They often use straw bales in fall decorations, and then are delighted to sell these props cheaply right after Halloween. Save the straw till the following spring in a garage or garden shed—it won't deteriorate—and then use it as a mulch in your vegetable garden.

Ongoing Plant Care 5

TAKE A TOUR OF YOUR GARDEN

Unfortunately, there is no such thing as a "no maintenance" garden. Some garden designers like to perpetuate the myth that there are certain, easy-to-grow plants that don't require any care at all. You just plunk them in the ground, then leave them to fend for themselves. Period. But the unshakeable truth is that all plants require some maintenance to varying degrees.

However, caring for a garden doesn't have to be a time-consuming, unpleasant chore. The trick to making sure plants stay in good shape (and require less attention) is to keep an eye on them. There's an old Chinese saying—"The best manure is the gardener's foot"—and it's very apt. If you walk around your garden regularly, inspecting everything, it's easy to assess how well plants are doing in their current locations, and to pinpoint specific problems and areas that need attention before they become major crises.

The best way to do this is to get into the habit of taking a leisurely morning (or evening) stroll around the garden. For many gardeners, this is the highlight of their day. It's fun, not work, to stop and examine plants, sniff their fragrances, pull out the odd weed, tidy up messy foliage, and keep an eye out for troublesome pests. And the more time you spend doing this, the more you'll become aware of the gradual changes taking place—buds starting to swell, blooms fading, a clematis vine getting out of control on the fence. It will also make you familiar with the areas that receive the most sun or shade, how the sun gradually shifts its position throughout the growing

Left: Regular inspection is the best way to keep plants in good shape. Tour your garden every morning or evening. Carry along a bucket, It's useful for weeds and spent blooms.

Balcony gardens need as much care as regular gardens. Potted plants can quickly outgrow containers and should be regularly checked.

season, and where the driest and dampest spots are in the garden—useful knowledge that helps determine what plants need to be moved—and where to place new purchases.

Carry a plastic pail or garden trug on these tours, and tuck a pair of scissors or little gardening shears in your pocket. The pail is useful for yanked-out weeds, and with the scissors you can deadhead spent flowers promptly. (When we have to go back to the house or garage to get tools, it's easy to put off these little maintenance tasks.) Some gardeners also stroll around with a notebook and scribble reminders to themselves about garden tasks that need to be done.

Inspect your garden regularly in the winter too. After a fall of wet snow, go outside and knock the snow off trees and shrubs. Check if there are any broken branches and cut off damaged bits cleanly with sharp pruners. If the wind has blown protective mulch away from plants, put these coverings back in place. And when spring comes, if bulbs and early perennials like hellebores are struggling to reach the light through layers of leaves, give them a helping hand by pulling back the leaves.

THE TOUCHY ISSUE OF WATERING

In our environmentally conscious times, watering has become a controversial topic. There's a growing body of opinion that insists we shouldn't be "wasting" water on our plants and lawns. Many gardeners now maintain that it's smarter (and kinder to the environment) to create gardens that can survive without water administered by humans.

These folks do make a valid point. It's also true that watering gardens regularly has always tended to be an urban and suburban habit. In rural areas, most residents don't water their gardens at all. They have the attitude that if their plants survive, fine. If they don't, they'll find something else to grow instead. The pros and cons on this touchy issue can be debated endlessly. However, it's undeniable that some plants, particularly annuals, benefit from regular liquid refreshment, particularly in the early stages of growth. Also, there are right and wrong ways to go about administering that precious water.

Wise Ways to Water

- Water deeply. Watering cans, strategically placed in gardens, certainly look attractive. But their use is best restricted to container-grown plants, seedlings, and newly planted annuals and perennials because most cans simply don't deliver enough water to established plants. Giving these plants a dribble of water from a watering can is, in fact, a self-defeating exercise because shallow watering encourages shallow, feeble roots that depend on more water. Most flowers, foliage plants, and shrubs planted in gardens prefer a good, deep drink once a week from the garden hose. (Roses are particularly partial to this.) Use a sprinkler and sink an old shallow tin can into an unobtrusive spot at ground level. When the water in the can reaches 1 in (2.5 cm) deep, you've watered enough. (With roses, train the hose on the roots, and avoid sprinkling the leaves, as this encourages mildew and black spot.)

- Water early in the morning if you can or, failing that, early evening. During the heat of the day, most water will evaporate (although the often-repeated claim that the sun acts as a magnifying glass on droplets of water, burning holes in plant leaves, is greatly exaggerated). Avoid the popular habit of standing outside after supper and training a garden hose on a flower bed or lawn. If foliage gets a blast of cold water at night, and it doesn't have a chance to dry off, plants are more prone to fungal diseases and attacks by pests such as slugs.

- Get a rain barrel (many municipalities now have programs to assist homeowners in buying them) and divert eavestrough runoff into the barrel. Failing that, a big plastic garbage can filled daily in spring with water from the hose is a good substitute. Dip watering cans into the barrel or bin and use the contents on delicate seedlings or just-planted annuals and perennials in both flower beds and containers. Room-temperature water is preferable for new plantings because it doesn't shock them the way icy water from a garden hose does. But don't leave undisturbed water sitting in an uncovered container. Once hot weather arrives, it's a perfect breeding ground for mosquitoes. You need to disturb the water's surface by dipping into the barrel or bin every day or two to prevent bugs from breeding there.

- Don't wait to water, as many gardeners do, until plants get "the droops." Wilting plants are much more susceptible to insects and diseases. (Some popular plants, such as perennial *Ligularia dentata* and annual African daisies *Osteospermum*, have an annoying habit of habitually wilting in the midday sunshine. These need to be kept permanently moist to prevent the droops.)

How to Cut Down on Watering

- Invest in a below-ground irrigation system for flower beds. More and more gardeners are doing this because they deliver water to where it's needed most (to the roots), they don't waste a drop (unlike sprinklers, which often end up watering driveways and patios as well), and water can be dispensed slowly, drip by drip. The drawback is that these systems are expensive and they must usually be installed before you plant anything, so they're out for gardens that are already established.

- Use a soaker hose, which is cheaper than a below-ground irrigation system. It works pretty well, but it can be difficult to weave the lengths of hose around established plants. Be sure to remove soaker hoses from the garden in fall and store them in a dry place. If left in the garden, they will retain water and crack.

- Strive to make your soil as moisture retentive as possible (see page 44) and use mulch around the base of plants (see page 75).

- Group plants closely together, without a lot of space between them. This helps to conserve moisture. Also, in new subdivisions, consider planting a leafy tree to provide some shade in the afternoons.

- Grow drought-tolerant plants. There are many to choose from now and most thrive anywhere, provided they receive plenty of sunshine. These include annual sunflowers, wormwood *Artemisia*, butterfly weed *Asclepias tuberosa*, California poppies *Escholzia californica*, coreopsis, cosmos, euphorbias, gaillardias, hens and chicks *Sempervivum*, *Liatris*, lambs' ears *Stachys lanata*, lavender, marigolds (both *Calendula* and *Tagetes* kinds), *Portulaca*, *Sedums*, yarrows, zinnias, some hardy geraniums, and many ornamental grasses.

Sunflowers are drought tolerant and look striking, but need lots of space.

- Don't expect perfection. Accept that plantings may dry out during hot spells and look less than their best. Homeowners who insist on immaculate lawns and flower beds are usually the ones who are most prone to waste water.

WEEDING: THE INESCAPABLE TASK

Many gardening experts like to pretend that weeds don't exist for they rarely mention them amid all the glowing suggestions about desirable plants to grow. There's also a growing body of eco-evangelist types who like to claim that every plant on the planet has some useful purpose and should therefore be treated with reverence, not contempt. Unfortunately, this wishful thinking is simply not true and, in some instances, it's counterproductive. Some of Mother Nature's weedier creations are actually a colossal nuisance to everyone from gardeners to farmers because they rob other plants of nutrients, moisture, and light. Once established, they can be very difficult to control because they either scatter seed everywhere or their roots are impossible to eradicate completely.

Small city gardens are unlikely to be bothered unduly by weedy interlopers, provided you keep a close eye out for them. Even so, a few undesirables can creep in. There are two basic kinds of weeds.

Annual Weeds

These come up in spring, live one summer, and then die. That's the good news. The bad news is that they frequently produce literally millions of seeds, so there are always more of them ready and waiting to spring into action and make a nuisance of themselves the following year. However, most annual weeds don't develop large root systems, so they are relatively easy to pull out. The worst of the weedy troublemakers in gardens are likely to be:

- Annual sow thistles
- Bedstraw *Galium mollugo*
- Chickweed
- Common groundsel
- Garlic mustard *Alliaria petiolata*
- Ground ivy
- Pigweed
- Purslane
- Ragweed
- Some kinds of grasses
- Wild cucumber vine *Echinocystis lobata* (This is an increasingly rampant vine that used to be restricted to agricultural areas, but is now making its presence felt in urban areas. It produces curious, spiky fruits that some gardeners find so appealing, they encourage the vine to multiply. Do not do this!)

Perennial and Biennial Weeds

These are usually more problematic than annual weeds because they live from year to year, often developing complex root systems that survive the toughest of winters. The most familiar are dandelions, which have fleshy taproots that can grow as fat—and much longer than—carrots. Dandelions are mostly considered a nuisance in lawns, but they can be big hassles in flower beds too because once those roots get a grip, they can be very difficult to dig out completely.

Perennial and biennial weedy menaces to gardeners include:

- Bindweed (The spaghetti-like underpinnings of this Boston strangler of plants often become entangled with the roots of other plants. They can also extend an amazing 30 ft (9 m) into the ground.)
- Burdock

- Common milkweed *Asclepias tuberosa*
- Dandelions
- Docks
- Dog-strangling vine *Vincetoxicum* spp.
- Some perennial grasses
- Quack grass and Bermuda grass
- Queen Anne's lace *Daucus carota*

However, in small city gardens, perennial weeds are unlikely to get a grip, provided they are removed promptly.

GETTING RID OF WEEDS

The best time to weed a garden is right after it rains. The soil should be damp, but not muddy. Weeds are much easier to pull or dig out if the ground isn't dry and caked. This is particularly true of clayey soils. Avoid stepping on the soil as you weed.

Weeding early in springtime is also important because that's when all plants put on their growth spurt. As Shakespeare once observed: "Now 'tis the spring and weeds are shallow-rooted: Suffer them now and they'll o'ergrow the garden." Yank undesirable plants out firmly before they become established and gain a hold. If removal is left till later in the season, annual weeds should be dispatched before they flower and have a chance to set seed. (Flowers are often tiny and virtually unnoticeable on many weeds, but look out for them.) In large areas, use a hoe with a sharp blade to snap off emerging weeds, although bear in mind that these tools leave the weed roots behind. For small gardens, there's a useful hand tool called the Garden Bandit, which does the same thing. A little hand fork for scratching at the soil, to loosen weeds, is also useful.

With perennial weeds, it's vital to remove every piece of root, which is often an impossible task by simply yanking with your hands. For dandelions, use a long spade (dig down deep) or a special dandelion digger. There are many on the market now. The best have big teeth at the end of their long handles, and these clamp down on the dandelion roots as you pull them out. Pointed hand trowels (with metal, not plastic, blades) are also useful for digging out perennial weeds. Also, look for a wickedly pointed tool from China called a Ho Mi digger, which is a boon in hooking out tough roots.

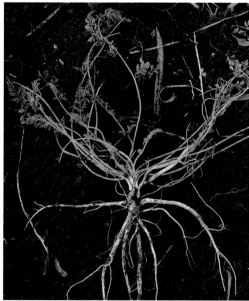

Top left: Pigweed
Middle left: Ground ivy
Bottom left: Bindweed in a yew bush
Top: Dog-strangling vine
Above: Queen Anne's Lace roots

A kettle of boiling water poured into the crown of many broadleaf weeds will kill them. Be careful not to spill the water on your feet.

Don't add weedy plants to a compost pile or bin unless the compost will be heated to a high temperature. Although it's desirable to recycle our garden debris, putting weeds in a plastic bag and leaving them out for the garbage collectors is preferable because most annual weeds contain some seeds, which will be transferred with the compost back into the garden. The remains of many perennial weed roots are also stubbornly tenacious. Even tiny, hacked-up pieces can hang on and send new roots, so it's best to banish these beasties from the garden entirely.

Remedies for Weeds

Organic

- Pour a kettle of boiling water over the offenders. This works well with weeds that grow in cracks between paving stones.

- A spray bottle containing vinegar is a surprisingly good weapon against weeds. However, don't use the kind that's sprinkled on french fries, because it's not strong enough. Try to locate a source of acetic acid (the scientific name for vinegar) in a 20 percent solution. Regular white vinegar contains only from three to five percent acetic acid. (The rest is water). Acetic acid is sold at agricultural and janitorial supply stores, as well as some hardware, automotive, and drug stores. If you can't find it, try Scott's Eco Sense Weed Control Spray, whose principal ingredient is acetic acid.

> ### ALERT!
> *Weeding ground covers*
>
> Swaths of low-growing plants (called ground covers) such as *Pachysandra*, *Vinca minor*, sweet woodruff, bugleweed, and ivies are increasingly being used in gardens to replace lawns. However, ground covers can wind up being as much trouble as a lawn because undesirable weedy plants keep forcing their way up among the desirable foliage. If you want to cover an area in a ground cover, be sure to dig out all weedy plants before planting. It may be necessary to kill off lurking weeds with an application of the herbicide Roundup. Follow instructions on the bottle. Repeat if necessary after fourteen days. Then plant fourteen days after that.

Alternatively, this homemade concoction may work: 4 cups (1 L) of household vinegar, 1 cup (250 ml) of lemon juice, 1/4 cup (60 mL) of salt and a few squirts of dish liquid. Put this in a plastic spray bottle and squirt directly on leaves. Most weeds will wither and die within days, although those with deep roots may require several applications. Desirable garden plants aren't damaged by acetic acid or salt because both get washed away by the rain. But don't spray this stuff on your good plants and don't use it too frequently. You can also apply it using the "double glove" trick. (See page 90.)

Chemical

Most home gardeners will never need to resort to chemical herbicides to eradicate weeds. If you do find it necessary because of a serious infestation in a large area of the garden, pick a product that has little or no residual toxicity. The safest on the market right now is glyphosate, usually sold under the brand name Roundup. (Other brands are Glifonox, Glycel, Kleenup, Rodeo, Rondo, and Vision.)

Roundup is a postemergent herbicide, which means that it kills over fifty types of grasses, vines, and broadleaf weeds *after* they have grown. Usually, it is diluted with water and sprayed onto offending plants. It then translocates, meaning that it goes down into the soil and attacks the weeds' roots.

The most effective time of year to use Roundup is just before weeds come into flower. It is less effective in early spring. With particularly tough weeds, such as bindweed and old dandelions, paint undiluted Roundup directly on leaves with an old paintbrush.

Avoid hitting desirable plants with the spray (because it will kill everything) and keep kids and pets away from treated areas for at least six hours. Don't use any chemical product when it's windy.

The "double glove" trick works on some weeds. Spray herbicide into the palm and grasp the offender firmly.

If you're worried about hitting the wrong plants when you spray, try the "double glove" trick. Put on a rubber kitchen glove, then pull a cotton glove over the top of it. Squirt some Roundup (or preferably acetic acid) into the palm of the glove, then squeeze the offending weeds. This is less risky than spraying, because there's no danger of the herbicide drifting in the wind and you can target certain plants. However, be sure to throw the cotton gloves away after use (they're sold cheaply in bundles at hardware stores) and don't use any chemical herbicides more than once a year.

Controlling Weeds with Corn Gluten

Corn gluten is a new, environmentally friendly breakthrough. It's caused quite a stir in the gardening world because this is the first weed killer that's 100 percent natural and harmless (pet food contains large quantities of corn gluten). It also adds a useful dollop of nitrogen to the soil. This product is beneficial mainly on lawns, but it can be used around transplanted vegetables, flowers, and shrubs to stop weeds from coming up.

Corn gluten works by zapping the roots of *sprouting* seeds. This is important to remember. It will not kill an established dandelion plant, but it will prevent that plant from producing more dandelions. Other weedy plants that can be stopped in their tracks by corn gluten include crabgrass, curly dock, black nightshade, purslane, lambs quarters, quack grass, black medic, red root pigweed, bedstraw, and creeping bent grass.

Timing is critical. Corn gluten must be applied in early spring, three to five weeks before weeds sprout. (As a cue, some gardeners wait till their crocus and early daffodils or forsythia start blooming.) An additional application in early fall, when temperatures are turning cooler, may inhibit seeds produced that summer, but springtime application is by far the best.

Sold mostly in pellet form in 5-lb (2.25-kg) bags, corn gluten should be spread evenly on lawns and around plants, and can be sprinkled in cracks on driveways and sidewalks too. It needs water to work, so apply before a rain or water in thoroughly. Then allow the area to remain dry for several days because dryness will help weed seedlings to shrivel up and die. (If it rains heavily, you may need to make another application.) While effective, corn gluten can take a couple of seasons to have a noticeable impact on weeds.

This product is more expensive than chemical herbicides and can be hard to find at garden centers. Stores selling environmentally friendly products sometimes have it.

Smothering (or Solarization)

Cover a weedy, neglected area of the garden with old pieces of thick carpet or sheets of black plastic for a season. This starves weeds of light, causing all but the most tenacious (such as bindweed) to give up. The following spring, dried-up, dead weed roots are much easier to remove and you can replant with desirable plants. It's possible do this at any time of year, but the smothering material should remain on the ground for at least two months, in order to kill the weeds properly.

Throughout the growing season, placing a mulch several inches thick around the base of desirable plants will help to stop weeds too. (See

Top: Dandelions have very long tap roots. Dig every piece out.
Above: Smothering an overgrown area with black plastic is easier than digging weeds out.

ONGOING PLANT CARE 91

Deadhead promptly. These snapdragons have been left too long.

page 75 for more on mulch.) So will massing plants close together. The idea is to fill the flower bed, leaving no room for weedy trespassers to sneak in and gain a foothold, but be careful of slugs multiplying under the mulch in wet weather. (See page 78.)

DEADHEADING

This rather gruesome term means removing dead or spent flowers from plants. Don't be afraid to do it. Deadhead and you'll get more blooms, better growth, neater-looking plants, and often longer blooming periods. Most gardeners actually find deadheading enjoyable because it's relaxing and provides an opportunity to examine plants closely.

Use kitchen scissors or small pruning shears with pointed, not rounded, ends to deadhead. They should have comfortable handles, preferably covered in plastic. With flower stems that snap off easily, simply use your fingers. Wear close-fitting gloves if you're doing a lot of deadheading with scissors or shears because hands can get sore.

Deadhead promptly once flowers falter and fade. Don't wait until they've turned brown and shriveled and seed heads are starting to form. The whole idea is to "trick" plants into blooming again—and they won't if they're allowed to set seed. Seed production also saps the plant's energy, often making it messy and straggly. A deadheading tour every morning or evening during the growing season is a good idea—and fun.

Which flowers to deadhead? And how? And do you just nip off the flower heads or prune a long way down the stem? The latter is the most frequently asked question about deadheading. Here are a few tips:

Snap tulip heads off when they have opened up and look like this.

- *Spring bulbs:* Snap off spent flower heads of tulips, leaving the stems, but leave narcissi standing. Pull out stems and leaves only after they've withered and turned brown. (This can take up to two months.) Don't deadhead small bulbs such as scillas if you want them to multiply in the garden. Leave spent alliums standing because they highly look decorative after going to seed.

- *Most perennials:* Deadhead these down to a lateral bud, flower, or leaf. This means you make your cut just above where new growth is starting on the stem. With flowers that have branching stems (such as daylilies *Hemerocallis*) be careful! Make sure you snip only the spent flower heads, not the new buds beside them. (These buds are often uncomfortably close together, which is why you need pointed shears to get between the buds.)

Some perennials that particularly benefit from deadheading are bee balm *Monarda didyma*, butterfly weed *Asclepias tuberosa*, most campanulas, *Penstemon barbatus*, most phloxes, salvias, shasta daisies, veronicas, and yarrows.

- *Perennials with long, single stems:* These include heucheras and hostas. Cut off the flower stem at ground level after blooming. (If you don't like the look of flowers on these foliage plants, remove the stems in early summer.) An exception is bearded iris. Since new flowers develop down the stem, simply remove flowers when they shrivel and go brown. Don't cut down the entire stem until all flowering has finished.

- *Perennials with lots of small flowers, such as hardy geraniums:* Don't deadhead (it's too tedious), but many benefit from shearing back (see page 94).

- *Most annuals:* Deadhead regularly for better bloom. It's particularly good for cosmos, fuchsias, geraniums (pelargoniums), all kinds of Nicotiana, petunias, osteospermums, snapdragons (snip off the entire stalk when seeds start to form), and verbenas.

Top: Gaillardia will bloom profusely without deadheading.
Above: Leave seedheads on sunflowers as birds love them.

Spent peony blooms look messy. Deadhead when their petals turn brown.

- *Annual herbs such as basil:* Deadhead flowers the moment they start forming, or leaf production will be drastically reduced.

- *Perennial herbs such as chives, thyme, oregano, parsley, and most mints:* Cut off flower stalks. You can also cut back the entire plant. (See below.)

- *Don't bother to deadhead:* Impatiens (flowers will keep on forming anyway); annual and perennial poppies (both have pretty seed heads); purple coneflowers *Echinacea* (new flower production is minimal and bees, birds, and butterflies love the seed heads); blanket flower *Gaillardia* (it will flower profusely in full sun and poor soil, whether deadheaded or not); sunflowers (all kinds are bird magnets). *Sedum* 'Autumn Joy', *Penstemon* 'Husker Red', black-eyed Susans, *Rudbeckia fulgida* 'Goldsturm' (and other *Rudbeckia* varieties). These all have delightful seed heads in winter.

- *Definitely deadhead these:* Peonies' spent flower heads look messy, like wadded-up old tissues, so cut their stems close to the ground; rampant self-seeders such as *Anchusa*, *Verbascums*, *Verbena bonariensis*, plume poppy *Macleaya cordata*, and most mallows. Let the flowers of these go to seed and you'll have "volunteers" popping up everywhere.

Remember: deadheading doesn't have to be done perfectly. Removing spent blooms promptly, so that more flowers can develop, is what matters. Carry a small basket or bucket while snipping, drop the dead flowers into it, and tip the castoffs on the compost heap.

SHEARING

Cutting back an entire plant takes courage, but some plants will bloom again beautifully if subjected to a drastic haircut. And fall bloomers will be less inclined to flop everywhere if you trim them back. Shearing is also much quicker than deadheading individual flowers (so it's useful in large areas), and it can be an effective way to reduce the height of certain perennials that are growing too tall or straggly. Also, if a plant looks dead, shearing may bring it back to life.

Use sharp garden shears, not power tools such as trimmers or hedge cutters, which tend to tear and injure plant stems. Don't shear on a hot day or if plants are growing in poor, dry soil. And in most instances, shearing is a no-no after midsummer (in most locations, July 15) if you want to promote flowers before frost.

Generally speaking, shear right after flowering. Open shears wide and make a clean cut through many stems at once. *Don't* hack at them. Leave at least 2 in (5 cm) of stem on plants. In most cases, don't cut right down to the ground. Look for a clump of leaves at the base of the stem and make your cut just above where new shoots and/or foliage are forming.

Don't shear new growth early in the spring or you may cut off flower buds.

Some plants that benefit from shearing:

- **Spring-flowering perennials** such as *Brunnera macrophylla*, cat mint *Nepeta*, evergreen candytuft *Iberis sempervivens*, Jacob's ladder *Polemonium caeruleum*, *Lamium maculatum*, mat-forming moss phlox *Phlox subulata*, *Pulmonarias*, and snow in summer *Cerastium tomentosum*.

- **Perennial herbs** such as chives, lavenders, mints, oreganos, and thymes. Harvest the first flush of leaves by shearing to promote an additional harvest later in the summer. (Cut off chive flowers right after blooming, and they'll often bloom twice more that season.)

- **Summer-flowering perennials** such as *Anchusa*, *Artemisias*, *Coreopsis*, *Euphorbia polychroma*, *Filipendipula*, foxgloves *Digitalis purpurea*, goldenrods *Solidago*, hardy geraniums (most kinds), *Santolina chamaecyparissus*, and *Thalictrum aquilegifolium*.

- **Annuals that are looking tired and leggy** such as *Anagallis*, *Bidens*, blue fan flower *Scaevola aemula*, coleus, cosmos, lobelia, nasturtiums, pansies, petunias, portulaca, verbenas, and zinnias can all be sheared back by half after blooming to promote new growth.

- **Ornamental grasses** such as ribbon grass *Phalaris arundinacea* var. 'Picta' often wind up looking straggly and brown at their leaf tips by midsummer. Shearing them down to the ground at that time will promote a fresh green crop of leaves.

- **Late summer bloomers** such as Michaelmas daisies, chrysanthemums *Dendranthema*, and black-eyed Susans can be sheared back by one-third at the beginning of July if they get too tall. Then they'll be less likely to keel over when they come into flower.

Big, wide clumps of tall perennials—especially ornamental grasses—can be difficult and time consuming to shear back. Wind some coarse string—or even masking tape—around the clumps before shearing, and the job will be easier and quicker.

Shearing Back Before Winter

Should perennials be sheared back to tidy up the garden before freeze-up? Some experts say yes, while others say no. In very cold climates, leaving dead stems and foliage affords protection to plants (and helps trap snow, which acts as an insulator). But if you've had an aphid, earwig, or slug problem the previous summer, it's best to remove all dead foliage in fall. (See page 106.)

STAKING

Staking means providing an artificial means of support to stop plants from toppling over. With some plants, particularly tall ones, it's a good idea to stake, but don't go overboard. A garden full of sticks and twine propping up plants looks ugly and unnatural. In fact, if you're forced to keep doing this, they may be the wrong kinds of plants for that particular location. In gardens that receive a lot of shade from surrounding trees and buildings (a common problem in cities), many sun-loving flowers tend to get spindly, weak, and too leggy. When that happens, it's best to replace them with plants that will tolerate some shade and grow in a sturdier fashion.

When staking, the trick is to remember to do it early in the growing season. Plants should be preferably no more than 1 ft (.3 m) high. This lessens the chance of damaging their roots. But, equally important, the stakes can be hidden more easily and blended in with the plant's foliage as it develops. If you don't shove in a stake until a heavy flower is flopping over (as most of us do), its stem will have to be yanked back into an awkward position, and what everyone will notice first is not the flower, but the stake.

Garden centers sell many kinds of plant supports. The best are dark green or a natural woody color, so they harmonize with foliage. Inexpensive bamboo canes work well for plants that have light single stems with big blooms on top (such as Maltese cross *Lychnis chalcedonica*). For heavier tall plants with flower heads arranged along the stem (delphiniums, for example), sturdier poles are usually necessary. Modernistic metal trellises, placed in the back of a flower bed, are great for this. For clump-forming floppy plants, (such as shasta daisies and peonies), use grid supports. These are positioned over the clump early in spring, and their stems grow through the holes. By midsummer, the support becomes invisible under the foliage.

Tomato plants, which get very heavy by summer's end, require cages or something very solid to clamber up. If you use homemade stakes, make them of heavy pieces of untreated lumber and hammer them deeply into the ground.

It's possible to recycle bits and pieces from the garden as stakes. Many tree branches work surprisingly well as plant supports and are great if you like a country-style garden. Leave twiggy bits on the branches to cradle the flowers and keep them facing upwards. An old piece of wide-mesh wire netting, bent into a U-shape and placed over a clump of peonies (which regrettably always require staking) when peony stems are just starting to grow in spring, makes a great grid support.

Push stakes in at least a couple of inches away from plant stems—not right next to them—and then make sure they are anchored *firmly* into the ground. (Wielding a hammer is always necessary in clayey soils.) Avoid having the top of a stake sticking up above a plant, and use something unobtrusive to tie stems to the stakes. Wire covered in a green foamy material (sold in rolls at garden centers) is excellent for most plants because it cushions the stems. Thin garden twine or nylon string tends to cut stems, particularly on windy days when plants sway. Garbage bag twist ties aren't advisable for the same reason. If you must use old panty hose (it's certainly soft and pliable), cut it into thin strips, so there won't be an ugly wad of nylon to spoil the appearance of your plants.

Most peonies require staking to stop their heavy blooms flopping over.

Some Plants That Benefit from Staking

- Amaranth *Amaranthus caudatus* (tall varieties)
- Asters (but pinching back in early summer will minimize flopping)
- Black-eyed Susans *Rudbeckia*
- *Cleome*
- *Cosmos* (although they look attractive, cascading everywhere)
- Dahlias (large flowered)
- Delphiniums
- Gladioli
- Hollyhocks *Alcea rosea*
- Joe Pye weed *Eupatorium fistulosum*
- *Lupines*
- Mallows (tall varieties)
- Peonies
- *Phlox*
- Poppies (tall varieties)
- Purple coneflowers *Echinacea*
- Shasta daisies *Leucanthemum* x *superbum*
- Sunflowers
- Tomatoes
- *Verbascums*
- Zinnias

However, avoid staking as much as possible. Find other ways to support plants. Use trellises and freestanding obelisks. Screw hooks in walls or fences, then run loose loops of foam-covered wire from the hook to the plant stem. And in a big garden—or if you love the English cottage garden-style of gardening—take the laissez-faire approach. Let some tall and unruly plants sprawl where they please because it looks charming.

PINCHING

If you're confused by this term, don't be. Pinching is simply nipping out buds or little new leaves, usually to create more flowers or a bushier-looking plant. It's useful and easy (although few gardeners bother to do it) and, unlike many gardening chores, it doesn't require tools. To pinch, simply use fingers and a thumb.

Buds directly on top of stems can be pinched off, just above where new leaf nodes are forming. New growth (buds or foliage) that pops up in the crook between a main stem and a side shoot can also be removed. Generally speaking, branching plants benefit the most from pinching.

Some plants to pinch:

- *Annuals such as pansies, petunias, and snapdragons:* Pinch them often to stop leggy growth and promote more flowers.

- *The herb basil:* Keep pinching off flowers and leaves appearing in crooks of stems to keep the plant bushy.

- *Tomatoes:* Many experts recommend pinching out little new leaves that form between branching stems, but this actually has no effect on the health (or size) of tomato plants. However, nipping off the first flowers that appear early in the summer will result in a bigger (but later) tomato crop.

- *Perennials:* Perennials such as chrysanthemums *Dendranthema*, Russian sage *Perovskia atripicifolia*, and *Sedum* 'Autumn Joy' can be pinched often, until midsummer, to stop them from being floppy. But don't do it late in the season, or flowers won't form.

THINNING

Overcrowded plants invite mildew and provide hiding places for creepy-crawlies. If plants become too dense during the growing season, it's perfectly acceptable to thin them out. In most instances, this won't inflict damage. Wielding the pruning shears actually helps growth and enhances the appearance of plants.

Gardening books usually recommend thinning plants early in spring by pinching off one in three stems at ground level. The problem with this advice is that most gardeners don't notice that plants are too densely packed until they're growing like gangbusters in the middle of the season.

Thinning can actually be done at any time. When plants look too crowded for comfort, simply pull clumps gently apart and cut out weak stems lurking underneath the stronger ones. Use sharp shears and cut close to ground level. It's most effective with medium to tall plants, such as *Artemisias*, asters, *Coreopsis*, delphiniums, hollyhocks, *Monarda*, peonies, phlox, and *Sedums*.

Deleafing

Removing individual dead leaves enhances the appearance of plants. It can also deter slugs and snails, particularly if the dead foliage is clinging to the bottoms of stems. (Slugs love to use yellowing leaves that touch the ground around tomato plants as ladders to climb up.) In spring, many perennials whose leaves stay green over winter—such as *Bergenia*, *Geranium maccrorhizum*, hellebores, *Pachysandra*, and periwinkle *Vinca minor*—benefit from a bit of tidying. However, don't go crazy. It's easy to wind up giving plants a peculiar look by removing too many leaves.

DIVIDING PLANTS THAT HAVE GROWN TOO BIG

Many gardeners fear that they'll wreck plants by digging them up and dividing them, but in most cases, this doesn't happen. In fact, some perennials can get out of control remarkably quickly, particularly in small city gardens. Dividing these plants becomes a must if you want them to look good and stay healthy. It's also not a good idea to let bossy plants smother the mild-mannered ones. One common cause of fungal infections in gardens is plants that are overcrowded, without sufficient air moving between the clumps. In such instances, plants get spindly and weak and are more prone to disease, as well insect attack.

Generally speaking, plants that garden centers label as "vigorous" or "robust" or "invasive" are the ones that require division most often. There's an increasing number of these plants on sale now. (In fact, if your garden is small, think twice before buying them!) While many older gardening books recommend dividing most perennials "every three to five years," there's no particular time frame to observe. Division depends on the type of plant, the growing conditions, the climate you live in, and the size of your garden.

Signs That Plants Need Dividing

- They're spreading over the entire flower bed.

- Their roots are getting tangled up with other plants, and new shoots are popping up among those plants.

- In spring, new growth appears in a circle, but there's a bare patch in the center of the plants.

- They're not flowering as often as before, or flowers are becoming leggy and sparse.

- The soil surrounding the plant has become clumpy and hard, and new shoots are having difficulty in breaking through.

When to Divide

Gardening gurus have different ideas on the "right" time, but they do agree on one point: Don't divide plants in summer, especially during hot spells or when a heat wave is threatened. The best time to divide is during cool weather in spring (before plants have put on lots of fresh new growth) or in fall. Many people opt for fall for two reasons. They're too busy with other gardening chores in spring, and after a summer of growth, plants that have spread and need dividing are more noticeable. If you do opt for fall, don't leave it too late in the year. A month before heavy frosts set in is best.

This clump of chives needs dividing. Use a sharp spade and cut cleanly.

Spring or fall, pick a day for dividing and replanting when rain is forecast because a good soak administered by Mother Nature seems to help moved plants settle in vastly better than water from the garden hose.

If you absolutely have to divide a plant in hot weather (or a friend offers you a chunk of something she's dug up), get it into the ground as soon as possible and keep the plant moist and shaded, with an upturned cardboard box, for a few days after replanting.

In any season, if you can't immediately plant, do not leave the plant clumps sitting in sunshine, with their roots exposed. Roots dry out extremely quickly! It's crucial to keep them moist and protected. Experts often recommend a procedure called "heeling in," which means digging a shallow, makeshift hole, laying the plant in it, and tossing some soil over the top. But if you don't have time for that (a familiar problem), simply put the plant in a completely shady spot, and place a thick layer of wetted newspaper (or some leaves) over it. Keep the paper moist until you're ready to plant it.

Plants that flower in spring are best divided right *after* they bloom. But summer and fall bloomers can be split up at any time in spring or early fall after they've finished flowering. The only exception is bearded iris, which should be divided in July or August.

It's easy to put off dividing plants. "The most difficult part is just getting around to doing it," points out garden writer Felder Rushing. "But when plants are divided," he notes, "it multiplies their beauty and the pleasure that they give."

How to Divide

Trim off foliage and flowers to a few inches from the ground. Dig around the plant with a spade that has a rounded edge, not a garden fork, which may chop into roots, damaging them. Go wide and deep, circling the plant. If you go too close, sections of roots are likely to be snapped off. And remember that plants will come out of the ground with less damage if the soil is damp, not dry as a bone. It's often a good idea to water thoroughly with the hose first.

Plants that have stringy roots (such as black-eyed Susans, hostas, hardy geraniums, campanulas, penstemons, purple coneflowers, periwinkle, and shasta daisies) are the easiest to divide. Often the soil will simply fall off them. Then you can pry the roots apart using fingers or a small garden tool.

However, tough customers with big root balls—astilbes, daylilies, peonies, Siberian iris, and most ornamental grasses—are another matter. Gardeners usually need the strength of Hercules to hack or pry apart these massive clumps. Use a sharpened, flat-edged spade, a lawn edger, a big kitchen knife, or a machete. Tip the root ball on its side and make a clean cut—wham!—right through the middle. If that's too daunting, try slicing off small portions from the sides of the root ball, or use two pitchforks, placed back to back, to pry the whole thing apart.

> ## ALERT!
> ### Clay and cold weather don't mix
>
> If you live in a very cold area of North America and have heavy clay soil, don't divide *any* perennials in fall. Wait until spring because subzero temperatures can heave these plants right out of the ground and zap their roots before they've had a chance to get established.

When dividing iris, be more careful. Cut out squashy or diseased-looking bits and plant only the tubers that look healthy and have roots emerging (or about to emerge) on their undersides.

Caring for Plants after Dividing Them

Dig a wide hole. Add some compost or other organic matter, fill the hole with water, and let the water soak away before planting. (See page 64.) If the root ball is dry and roots look tightly jammed (which often happens with old perennials), pry some soil off the root ball to loosen everything up, then splay the roots out around the hole with your hands. Refill, tamp down, and water well.

Put a protective mulch around the base (see page 75) and keep an eye on the plant for the next few weeks. Water regularly. It's important not to let newly divided plants dry out.

Many perennials won't flower the first year after being divided. And some plants take as long as three years to settle into their new homes.

VOLUNTEERS: A MIXED BLESSING IN MANY GARDENS

Gardeners have a name for plants that pop up in the garden without any help from humankind. They're "volunteers." Although it can be thrilling to discover that a certain plant has gone to seed and produced offspring on its own in our gardens, volunteers are actually a mixed blessing. Some can quickly get out of hand because they keep sending up seedlings everywhere—in flower beds, lawns, containers, and cracks between paving stones and pathways. You may find yourself constantly bending down to remove these enthusiastic newcomers.

Whether or not volunteers become too much of a good thing depends greatly on the climate, growing conditions, location, size of the garden, and the kind of gardener you are. Certain kinds of plants can be particularly problematic in sunny gardens in the South. Others are stimulated to set seed by good (or poor) soil, and still others will self-seed like mad even in shade. Keep prolific plants to a minimum by lopping off their flower heads before they go to seed, but don't banish them entirely unless you have to because they're often beautiful or, in the case of herbs, useful. If they become impossible to control, remove all established plants in the fall, then yank out the seedlings the moment they appear in spring. The seeds of some volunteers can remain viable in the soil for several years, so it may take time to eradicate them.

These plants are inclined to become a nuisance:

Herbs

- *Allium schoenoprasum,* chives
- *Anethum graveolens,* dill
- *Angelica archangelica,* angelica
- *Anthriscus cerefolium,* chervil
- *Coriandrum sativum,* coriander
- *Melissa oficinalis,* lemon balm
- *Nepeta cataria,* catnip
- *Origano,* oregano
- *Rumex,* sorrel
- *Tanacetum parthenium,* feverfew
- *Valeriana,* valeriana

Flowers

- *Alcea rosea,* hollyhocks
- *Alchemilla mollis,* lady's mantle
- *Anchusa,* blue bugloss
- *Artemisia,* most kinds
- *Dianthus deltoides,* maiden pinks

- *Echinacea*, purple coneflowers
- *Hesperis matronalis*, dame's rocket
- *Impatiens glandulifera* Himalayan balsam
- *Lysimachia*, various kinds
- *Malva alcea*, hollyhock mallows
- *Myosotis*, perennial forget-me-not
- *Oenothera Missouriensis*, Ozark sundrops
- *Onopordon acanthium*, Scottish thistle
- *Papaver somniferum*, annual opium poppies
- *Rudbeckia* var. *sullivantii* 'Goldsturm', black-eyed Susans
- *Silybum marianum*, milk thistle
- *Verbascum*, most kinds
- *Viola*, common violets

Left: Himalayan balsam *Impatiens glandulifera* (here with *Nicotiana sylvestris*) is a rampant self-seeder. Above, foreground: Some mulleins *Verbascum* spread seeds too prolifically for small gardens.

New hybrid varieties of some of the above plants do not self-seed as readily as the original versions and, in some instances, do not reproduce themselves at all. For example, new white and orange varieties of purple coneflowers are not prolific seed producers. And a popular new *Verbascum* variety called 'Copper Rose' F1 hybrid is incapable of setting viable seed at all.

Plants That Arrive as Gifts

Gardeners hate throwing anything away, so sooner or later friends are bound to offer you plant divisions. They can be welcome. One of the greatest pleasures of gardening is sharing. However, check gifts carefully. Before you plant, remove any undesirable weeds clinging to the root ball. Don't introduce anything to your garden that looks diseased or buggy. And be wary of invasive plants that may be tangled up in the roots of a desirable plant. If you're offered something that's been growing in a ground cover called goutweed *Aegopodium podagraria*, say "No, thanks," immediately. Goutweed roots are the most tenacious on the planet. Once introduced to your garden, they will be around forever, and they are impossible to get rid of.

WHAT TO DO WHEN WINTER COMES

Opinions differ about the time-honored practice of "putting the garden to bed." Some folks prefer to cut all perennial plants down to the ground in fall and leave everything neat and tidy. Others maintain that denuding gardens before the onset of winter is unattractive and that in cold climates it can actually do more harm than good.

Both viewpoints have merit. Foliage on some plants (such as daylilies) should be cut off in fall because it gets mushy and moldy if left lying on the ground over winter. All kinds of foliage may also act as a haven for slugs and other overwintering garden pests. But many other decaying plants can be highly attractive in the winter landscape because their shapes, textures, and colors add interest at a time when everything else looks a bit bleak. As well, seed heads provide food for birds, and old foliage and stems can act as a nesting ground for desirable garden insects like butterflies and moths. A still-standing plant is also better equipped to face the ravages of winter because snow (the best insulating material there is) is less likely to blow away if it has somewhere to collect and settle.

Fall cleanup is ultimately a matter of personal choice. If you dislike the sight of last summer's plants flopping around (or have had a problem with any kind of insect the previous summer), follow the trim-and-tidy routine. If those garden leftovers don't bother you, cleanup can wait till spring. Whatever your course of action, fall is absolutely the best time of year to take stock of the garden. Make a tour. Examine everything carefully. Assess what worked and what didn't. Push clumps of plant leaves aside and look underneath them. Are the plants spreading too much? Have their stems become too cramped? Is sufficient water getting to the roots? Have the plants started to squeeze out neighboring plants or are they being squeezed out themselves? Decide what you need to move, divide, get rid of, and add. And, at the same time, take the trouble to dig out stubborn perennial weeds (like dandelions) in the declining days of the gardening season. Few gardeners like weeding in fall, but if you do, those annoying hangers-on won't be popping up again and causing problems the following spring.

Steps to Take in Fall

- Rake leaves off flower beds into heaps. (Careful! It's easy to pull loose-rooted plants like hardy geraniums right out the ground when raking.) Remove leaves from pathways where they may become slimy and slippery. Use them as mulch around borderline hardy plants. (See page 75.)

- Pull up and compost all low-growing annuals, such as impatiens and petunias. The right time is after the first frosts when such plants turn mushy and black. Taller annuals, such as rudbeckias, sunflowers, coreopsis, cosmos, and *Nicotiana sylvestris* can be left standing because birds love their seeds. Sprawling hollyhocks should be cut down and their foliage cleaned up. If the leaves are stippled with rust (see page 155), do not compost them.

- Cut down messy perennials when they go into dormancy after the first frosts. Leave 3 in (8 cm) of stem on most plant clumps. (Peonies, which produce their new shoots below the surface, can be cut down right to the ground.) If foliage and stems have mildew, remove them, but don't add them to the compost bin. Leave standing any plants that have attractive seed heads or foliage. Particularly appealing are penstemon 'Husker Red', purple coneflowers, black-eyed Susans, sedums, all kinds of sunflowers, and verbascums. Many of these plants are magnets for birds and will look glorious in fall, then quite passable for the rest of the winter.

Leave seedheads on decorative plants like *Rudbeckia hirta* as they provide food for birds.

If you want to dig up and move perennials, do it about a month before the first frosts. (See page 61.)

- Do not cut back lavender, which may be permanently damaged if pruned hard before winter. In cold climates, it's best to prune in spring. The same goes for sage, whose leaves stay green a remarkably long time. (Even in the far North, you can brush the snow off sage and pick leaves, still fresh, for the Christmas turkey stuffing!) Prune back mints, thymes, and oregano only if they're leggy and messy, as perennial herbs are less likely

Ornamental grasses can be left standing all winter, then sheared back in spring.

to experience winter dieback if foliage and flower heads remain on the plants until spring. Pull out basil (an annual herb), and compost it. (It goes black the moment frost hits.) Parsley, a biennial, can stay in the ground as is.

- Because they rustle pleasantly in winter winds, leave all perennial ornamental grasses standing unless they are flopping and getting in the way. Cut the foliage down to the ground in early spring instead. Leaves of some thick-bladed varieties, such as ribbon grass *Phalaris arundinacea* var. 'Picta', can be recycled at that time as summer mulch around plants. Big clumps of grass can be time consuming to cut down. Tie them loosely together with coarse string or duct tape first, then use garden shears. If clumps are huge, an electric hedge trimmer is less work.

- Don't cut down perennial vines in fall unless they're out of control. Most survive winter best if left standing. Spring is a better time to prune them. Hill up earth around their bases.

- Empty annual flowers from small containers. Scrub out the containers in a bucket of water with a bit of bleach added. Stack them in a shed or garage. Big planters can be left as is. Just pull out the dead annuals.

- Put a protective layer of mulch (leaves, evergreen branches, or composted manure) over container-grown perennials that you intend to leave outside. Borderline hardy ones such as hostas should be brought into a garage for the winter. (The larger and deeper the container, the more chance the plant will survive out of doors.)

- In cold areas, coddle roses (except hardy shrub roses) by mounding up soil around their stems before freeze-up. Soil should be at least 6 in (15 cm) deep, then layer on some leaves or straw. (Or in January, cut up branches from the Christmas tree and lay those around the rose stems.) A rose collar, sold at garden centers, will keep soil in place. Alternatively, cut the bottom out of a flexible 2-gal (7.5-L) plastic pot, then slash one side open, top to bottom, and

ALERT!
Clay pots crack in winter

Clay pots are more likely to remain intact if you raise them off the ground on little feet or a pile of sticks or stones. The key is to provide good drainage. If there's somewhere for melting snow and ice to run off, the water won't puddle, then freeze solid, expand, and crack the pot. Look, too, for clay containers with thick sides and bottoms that are fired at very high temperatures. Potters often make them (make sure the pots have drainage holes). Flimsy clay pots, imported from tropical climates, are less reliable. And if you're doubtful about prize perennials planted in clay containers (or winterproof resin ones) surviving the winter outdoors, haul the pots into an unheated garage for the winter months. Container-grown hostas are particularly good at adapting to this treatment as long as they're watered occasionally.

wrap the opened-up pot around the rose stem, then fill with soil. If roses have experienced a lot of vegetative growth over the summer, remove lower leaves and all debris from around the base of the plant because this decaying material encourages insects and slugs.

- Mulch all flower-bed plants that are borderline hardy or recently planted (see page 61).

- Bring container-grown tropical plants indoors before frosts. Most can be pruned back when they come inside. Reduce watering for the winter.

- Most houseplants benefit from a bath before coming back inside. Plunge them, pots and all, into a bucket of water with a few squirts of a mild dishwashing detergent (like Dove) added. Immerse the pot till bubbles appear, then turn it upside down and dunk the leaves too. Rinse off with the garden hose or a bucket of clean water. This kills insect eggs laid in the soil during the plants' summer sojourn outside.

Wash houseplants with insecticidal soap after summer outdoors. Rinse under the tap or in a bucket.

If containers are too big for the dunk treatment, wash off leaves with a sponge, then tip the remaining soapy water through the pot's soil and rinse. You can also try putting containers in the bathtub and turning the shower on for five minutes.

- Pile leaves into a compost bin or heap if you don't intend to use them as mulch, or compost in black plastic bags. (See page 76.)

- Water perennials well in fall. Position the garden hose over their roots and let water trickle out slowly for a couple of hours. This is particularly valuable if the summer has been dry or if you've recently put the plants in.

- Do not fertilize anything.

WHEN SPRING ROLLS AROUND AGAIN

Do a tour to see what made it through the winter, but don't be in too much of a hurry to tidy everything up or dig up plants that look dead. Peel back mulch carefully. Wait till shoots are starting to pop up on early perennials and bulbs before getting out garden tools. It's very easy to wreck prize plants by raking up winter detritus too forcefully and by digging around in the soil before you can recognize where all your plants are located. Remember, too, that some perennials and bulbs take a long while to send up new growth. Let them do it according to their own schedule, not yours.

WHEN ROSES WON'T PRODUCE FLOWERS

This phenomenon is called "rose blindness," which happens with some types of roses. They will flower poorly, year after year, in spite of good care and healthy growth. Here are some tips:

- An occasional blind shoot with an empty flower case can usually be traced to frost damage.

- If blind shoots happen all the time, the problem may be too much shade or a soil that's not fertile enough.

- Cut back blind shoots by half to stimulate new growth. Improve the growing area by working in a rose fertilizer and placing a mulch around the base. Check that the rose is receiving sufficient sunlight. Most need at least six hours a day. If the rose used to flower well, the problem is often that surrounding trees or shrubs have grown bigger and are now blocking out the sun.

- In climbing roses (for example, 'New Dawn') blindness may be caused by a buildup of old, woody growth. Prune this out drastically, and see if flowering improves the following season.

ALERT!
Avoid rock gardens

They may be in vogue, but they're also notoriously difficult to weed. Roots of undesirable plants often become wedged under the rocks or they get entangled with the underpinnings of desirable plants, and you end up having to remove rocks to get them out. In fact, in order to have a weed-free rock garden, it's advisable to spray the area with a herbicide like Roundup (twice, at two-week intervals) before planting anything. But steer clear of this type of gardening if you want low maintenance because rock gardens are definitely more work than regular flower beds.

Protecting Plants from Problems

6

HOW TO CULTIVATE GOOD GARDENING HABITS

A healthy garden—untroubled by bugs and diseases and full of flourishing plants that all perform the way they're supposed to—is the dream of every gardener. And many experts insist that this Utopian ideal is easy to achieve as long as we respect the environment, buy the right kind of plants, and follow certain rules about growing things. The truth is, though, no garden is likely to be perfect 100 percent of the time. Mother Nature can place challenges in the paths of mere mortals, and while there are countless joys and successes to be had in any garden, inevitably a few disappointments and failures will be part of the package. Some plants work. Some plants just don't. Period. Even with the best of intentions, they may fail to thrive due to circumstances beyond our control. However, the way to minimize the disappointments is to cultivate good gardening habits. Here are some tips.

Left: Aim for a diversity of plants in your garden and provide good air flow. If bugs or diseases hit, act promptly. But don't expect perfection, because no garden is ever perfect.

Buy Insect- and Disease-Resistant Varieties of Plants

Always look for these because growers are constantly experimenting with plants, and coming up with new cultivars that will battle old, familiar problems. For instance, many garden centers now stock hollyhocks that are resistant to rust, roses that are not afflicted by black spot, phlox and *Monarda* varieties whose stalks won't inevitably be covered in mildew, and tomato seedlings that are immune to the dreaded damping-off disease. Check labels when buying. However, bear in mind too that, with some species of plants, original kinds tend to be tougher and less prone to problems than the newer cultivars. In fact, when a new variety is launched, it's a good idea to wait a year or two to see how well it performs for other gardeners before buying it yourself—much in the same way that automobile experts recommend that we delay buying a new vehicle until its second or third year of production.

Stick to Plants That Suit the Local Climate

It's not worth buying something that won't do well in your area, however tempting it may look in the garden center. (See page 15.)

Embrace Cultural Diversity

Mix things up. With lots of different types of plants in your garden, minor problems are less likely to escalate into serious ones.

In a monoculture (that is, a large area of one type of plant), pests and diseases settle on one host plant, then move on and have a field day with all the rest. (It's why those traditional rose gardens containing nothing but rows of roses are so prone to diseases like black spot.) Plant an assortment of perennials. And if you're growing vegetables or annual bedding plants, don't keep planting the same things in the same places, year after year.

Provide Good Air Flow around Plants

This is a huge, underestimated factor in the spread of plant diseases, particularly fungal infections. Lack of air circulation encourages insects, too, particularly aphids, because overcrowding weakens plants, and encourages the little nasties to get a grip. We're all guilty of trying to pack too much into our gardens—there's so much out there to buy, right? But the less-is-more philosophy does pay dividends. Fewer, carefully chosen plants look better than a mishmash of all different plants anyway.

Pay Attention to Weather Forecasts

Increasingly, weather patterns are unpredictable all over North America. It's cold when it should be hot, and vice versa. We may get late frosts in spring and frosts too early in fall. There are searing droughts or torrential rain, coupled with floods, in areas that never used to experience such things. Unexpected gales may destroy trees. Take warnings about extreme weather seriously and make the effort to protect your plants before it hits. Also take changing worldwide weather patterns into account when buying plants. The ones that once were deemed to be a sure thing in your area may no longer be the best choices. But on the plus side, with global warming, you may also be able to grow things that once weren't winter hardy in your area but now are.

Act Fast When Bugs and Diseases Hit

If a plant is looking sickly, don't dither. It's important to stop insect problems before they become full-blown crises. (Most of us wait too long.) But avoid simply rushing out to the garden center and looking for a product that will "zap" the problem. Although there's a huge number of offerings, both chemical and organic, on the market—all of them catering to the desire of anxious consumers to banish everything from earwigs to aphids with a few quick squirts— take the time to analyze what's happening to the plant. Try to find an inexpensive, environmentally friendly, homemade solution first. And become knowledgeable about pests and their lifecycles because then it's easier to control them.

When you think the problem is some kind of bug, pull off some leaves and examine them (checking both top and undersides) carefully. Also check the flower petals and stems. Use a magnifying glass if you can't seem to detect anything. Some insects like aphids are cunning little critters, barely visible to the naked eye. Look up the bug in a gardening book and when there's nothing that resembles yours, cut off an affected section of the plant and take it to a garden center for advice.

ALERT!
Don't spread diseases and bugs

Garden center staff are usually more than willing to try to identify the cause of an ailing plant. Take a cutting along to the center and ask somebody. But *always* place the cutting in a plastic bag that's sealed with a twist tie before you leave home because insect infestations and plant diseases can spread easily.

Plant diseases are usually trickier to resolve than insect problems because in some instances, the cause is a systemic virus that affects the whole plant and there's no cure. When you notice a plant that looks diseased, remove the affected parts right away to prevent the disease from spreading. If an entire perennial looks wizened or moldy or blotched with something peculiar, cut it right down to within a few inches of the ground and throw out the detritus, sealed in a plastic bag. With diseased or ailing annual flowers and vegetables, it's often not worth keeping them around. Damping-off disease, for instance, is a devastating condition that can spread like wildfire through seedlings. Get rid of the affected ones right away because they can quickly infect all the others.

And don't hang on to any plant—annual or perennial—that is persistently bothered by pests or diseases. It's not worth the hassle. There are plenty of other wonderful things to grow.

WHAT IPM MEANS AND HOW IT AFFECTS YOU

Integrated Pest Management is a term that's being used more and more in the gardening world. You'll hear it a lot at garden centers and from lawn care specialists and landscapers, which is a good thing because IPM stands for a new, more environmentally friendly approach to tackling garden pests and diseases. Instead of simply reaching for a chemical, professionals in the horticultural industry are increasingly in agreement that problems need to be treated in a holistic way. That is, if you wipe out a bug or disease with a particular product, will you at the same time be wiping out something beneficial? Would it be better to do something else instead? What's causing the problem to occur in the first place? And is there a less drastic treatment, or combination of treatments, that can be introduced to control the problem?

With IPM, professionals use all the resources at their disposal—biological, physical, and chemical (both natural and synthetic)—to make things work. What this new attitude means for the rest of us is this: A lawn care company that once blanketed everything in a noxious herbicide/pesticide concoction may now switch to growing a different type of grass that crowds out weeds and is better suited to the local climate. Then it may add organic matter to bare areas in the lawn, and only spot treat a few of the toughest weeds with a herbicide. If you'd like to know what methods the lawn care trucks that visit your neighborhood are using, ask if they've adopted IPM, and if they say no, call up the company boss and ask why not because this is a practical approach to a controversial issue.

IPM makes sense for home gardeners too. When bug infestations or plant diseases occur, see if you can figure out why they happened before reaching for a chemical. And then look for a solution that will not upset the garden's ecosystem.

BENEFICIAL BUGS IN THE GARDEN

There are lots and it's a good idea to learn what they look like because they help keep the "baddies" in check. You can order these insects by mail from companies that specialize in biological controls. However, it's better to create the kind of environment that beneficial bugs like because then they may show up on their own. This can be tricky because specific bugs are attracted to specific plants but, as a general rule, bugs like places in the garden where they can hide—grasses, vines, plants that have lots of foliage, and flowers to perch on. They also need a water source and do not tend to hang around in bare, open spaces, where it gets too hot. If you have a vegetable garden that's plagued by bugs, cultivate a stand of flowers, particularly wild flowers, next to it. Here are some bugs worth having as friends:

Green Lacewings

Green lacewings are great general predators because they eat all manner of pests—aphids, leafhoppers, mealybugs, scale, spider mites, thrips, and whiteflies, as well as the eggs of many caterpillars. As insects, they are fragile-looking and green, with small heads, large eyes, and transparent, lacey wings, and they live on pollen and nectar. It's the larval form, called the "aphid lion," that does all the beneficial munching. When bought by mail order, lacewings usually arrive as eggs, in grains of rice, or as larvae.

Ladybugs

Everyone knows what ladybugs look like: they're red with black spots and they look "cute." Aesthetics aside, they're a huge boon in gardens because they eat aphids (also Colorado potato beetle larvae, chinch bugs, thrips, mites, and others). One ladybug may consume over 5,000 aphids in its lifetime. Many people buy ladybugs in the thousands to release in their gardens, but this often doesn't work. They die or simply disappear, particularly if they're introduced to a climate that is different from where they came from. It is also possible to wind up with too many ladybugs—then they will keep hatching and flying around in annoying swarms. A better plan is create a natural population of ladybugs by growing plants they like. These include goldenrod, yarrows, morning glory vines, Queen Anne's lace, and nasturtiums. Learn to recognize ladybugs in their larval stage too. They're not very pretty—a bit like little black alligators with orange marks on their backs—and they cling to leaves. But don't squash them! Also be careful of a ladybug look-alike that is brown with (sometimes) orange spots and has been imported from Asia to control aphid populations in some jurisdictions. This little critter actually bites people—hard.

Nematodes

There are good and bad nematodes (see page 142). All are thread-like worms, which are active in the soil. The beneficial ones can be useful in controlling soil-dwelling larvae of insects, such as grubs in lawns, borers, cutworms, beetle grubs, fungus gnats, and root weevils. They do not attack plant roots as the "bad" nematodes do. You can buy the good kinds by mail order; then they must be mixed with water and sprayed on the soil because they are so microscopic. However, nematodes are tricky to introduce in some areas because they become inactive when the soil temperature dips below 55°F (13°C). If you live in a cold climate, look for a kind of nematode called *Steinernema feltiae*, which is cold tolerant and has been proven to last up to eighteen months in soil and lawns in northern New York state.

Parasitic Wasps

These are tiny, and there are many kinds with bewildering classifications and names. The best known—and one of the most useful for home gardeners—is the trichogramma wasp. It won't sting us, but it will deliver a death blow to a long list of insect pests, including aphids, cutworms, leaf rollers, cabbage loopers, mealybugs, scale, whiteflies, many moths, and various beetles. Parasitic wasps perform this dirty deed in a curious, roundabout way: the female lays her eggs in the pest's eggs, then when the wasp egg hatches out, it gobbles up the hapless host. You can actually see them do this on tomato hornworms. (See page 147.)

Praying Mantis

These are fun to have around. Kids love them because they look so *weird*. They are huge and green (or sometimes brown) with long folding legs and will pop up in expected places, usually toward the end of summer. But praying mantis are actually a bit overrated in the beneficial bugs department. While they eat soft-bodied insects such as aphid and leafhoppers, they don't consume vast amounts. They also, unfortunately, will dine on "good" insects too—like bees—and on each other. They are attracted to goldenrod and are very territorial. Once established in a certain place, they'll tend to stay there, and drive off other praying mantis or even eat them. You can sometimes see their egg clusters, which resemble blobs of papier-mâché, hanging among the fuzzy fronds of wild goldenrod in fall.

Spined Soldier Bug

This looks like a stinkbug and belongs to the same family. It's about $1/2$ in (1.2 cm) long, brownish with a shield-shaped back. "Soldier" is an appropriate name because this one likes to spear victims. Then it injects a paralyzing venom and sucks the life out of the helpless enemy, usually a caterpillar. These bugs are general predators of all kinds of caterpillars and are particularly useful in a vegetable garden.

DOES COMPANION PLANTING WORK?

Yes and no. There's no question that certain pests (insect and animal) seem to dislike certain plants and steer clear of them. But companion planting is an inexact science, and overexaggerated claims abound. In the final analysis, good gardening habits play a far bigger role in achieving a healthy, flourishing garden than planting a certain plant in close proximity to another plant. However, some may act as deterrents. Here are a few tips:

- Plant marigolds—the *Tagetes* kind, not the old-fashioned marigolds called *Calendula*—anywhere you want to keep bugs at bay. *Tagetes* have a very strong smell, which some people hate. Many kinds of insects and caterpillars do too. These marigolds are also said to make "baddy" nematodes move elsewhere in the soil. A few *Tagetes* planted among vegetables definitely helps deter bug infestations. Combine them also with annual flowers planted in containers and they may keep whiteflies and slugs away. Any variety of *Tagetes* seems to work, even the dainty *Tagetes tenufolia* 'Lemon Gem' and 'Tangerine Gem'. These two have yummy citrus scents and look far prettier than their mop-top, lurid orange cousins. Buy marigolds as started plants at the garden center in spring. They are also easy to start from seed indoors in early spring

- Most pests seem to hate onions and garlic (probably because of their smell). You will rarely see a slug sliming its way up an onion stalk, so include a few onions, chives, and garlic chives in a flower bed. Also plant garlic cloves randomly (in the fall) among plants. Besides being a bug deterrent, garlic can look very striking mixed with flowers as the stalks

The strong smell of marigolds (foreground) may keep pests away from vegetables.

Top: Garlic heads look surprisingly decorative and deter some pests. Above: Crown Imperial *Fritillaria imperialis* has a skunky smell that squirrels don't like.

may soar 5 ft (1.5 m) tall (in sun); they have fascinating pointed buds with casings that peel back to reveal flower heads as pretty as any ornamental allium. In a vegetable garden, include *lots* of onions among the other veggies.

- Many perennial herbs—hyssop, lemon balm, various mints, rosemary, sage, thymes, and wormwood—seem to deter insects and creepy-crawlies, so include them in flower beds. However, the opposite is true of the annual herb basil. This leafy plant seems to be a magnet for pests, particularly slugs and earwigs, and they will chomp basil plants to shreds very quickly if allowed to lurk undetected. (It is best to grow basil in containers on a deck, where you can keep an eye on it.)

- Bugs seem to dislike rhubarb, so don't restrict it to a vegetable garden. Include a clump in a sunny flower bed, where it looks surprisingly decorative. (You can also spray buggy plants with a solution of chopped-up rhubarb leaves that were boiled in water.)

- Plants with strong, earthy odors (as opposed to delicate perfumes) turn off some bugs. You'll rarely see any kind of insect or creepy-crawlie on *Geranium maccrorhizum*, a very useful ground cover that has a powerful smell, which some gardeners dislike too. Some insects, notably whiteflies, give the cold shoulder to nasturtiums, which also have a powerful smell, although earwigs and slugs have no such prejudices; they love nibbling nasturtium leaves. However, this strong-odor phenomenon is by no means universal. For example, tomato plant leaves have a strong smell, but a variety of pests will chomp like mad on them.

- Squirrels and tulips don't mix, as every bulb-loving gardener knows. To keep the bushy-tailed varmints away from your spring display, try *Fritillaria imperialis* or crown imperial, a strikingly handsome spring bulb that is planted at the same time as tulips. It exudes a very powerful skunky scent that has been proven to scare off some, but not all, the critters. (Success or failure probably depends on whether or not the squirrels are used to sharing their turf with skunks.) Buy *Fritillaria* bulbs in fall (they're big and expensive) and plant them on their sides because they have a tendency to fill with water and rot if planted upright. Choose a spot that's close to the tulips, but avoid planting near a deck or patio because that strong odor can truly knock your socks off. A small relative, *Fritillaria meleagris*, has no scent. The big "frits" tend to last only a couple of seasons in the garden.

THE DIFFERENT KINDS OF PESTICIDES SOLD AT GARDEN CENTERS

"Pesticides" is the catchall term now used to describe pest-control products, even though they may contain one or several of the following:

- Bactericides, which are designed for use on bacterial problems
- Herbicides, which kill weeds
- Fungicides, which tackle fungal problems
- Insecticides, which kill insects
- Miticides or acaricides, which control mite-like pests
- Molluscicides, which kill slugs and snails

ALERT!
Zap 'em with this spray

Most pests gag at the smell of garlic and they hate hot pepper. Grind up a bunch of garlic cloves in a blender (about half a dozen are fine) and add 3 cups (750 mL) of hot water and 1 tbsp (15 mL) of cayenne pepper. Blend, then leave this concoction to marinate for a couple of days. Strain through a coffee filter and pour into a misting bottle. Spray on plants. It's fairly effective with some soft-bodied insects, but must be repeated in serious infestations and after rain.

These products may be chemical (manufactured from synthetics), organic (derived from natural sources), or sometimes a combination of the two. The chemical kind are divided into several categories:

- Systemic pesticides, which get taken up into the plant and ingested by insects that suck or eat it
- Nonsystemic pesticides, which are not absorbed by the plant and work directly on pests
- Contact pesticides, which are designed to kill or paralyze the pest on contact
- Selective pesticides, which kill certain things that they touch, but do not affect others (such as beneficial insects)
- Nonselective pesticides, which kill everything they touch (and are absolutely the worst kind of pesticide to use)
- Baits, which may lure the pest into a trap and contain some kind of pest attractant

These products are sold as sprays, emulsions (which must be mixed with water), powders (stirred into water), and ready-to-use liquids, dusts, granules, and food baits. Whatever their formulation, they must, by law, carry labels containing a whole lot of information. This includes the type of chemical it is, the active and filler ingredients (listed by percentage of volume or weight), the safest time to apply the chemical, complete directions on how to use it, first-aid treatment, and, finally, what are called "precautionary statements." When you read these statements, be aware that:

- DANGER means the most toxic chemicals of all, which are extremely poisonous to humans and animals, and may be fatal if swallowed.

- WARNING means highly toxic, but not as deadly as those marked DANGER. This product can still kill you, but you need to ingest larger quantities.

- CAUTION means it's going to take more of this chemical to kill you or your pets than the other two, but it's still somewhat toxic.

Although they're generally lumped together by the public, some chemicals are much more toxic than others. Also be aware that reputable gardening companies, staffed by professionals, are, generally speaking, knowledgeable and responsible about the use of these products, and that home gardeners are the ones who cause many of the problems with pesticides because we don't bother to read the instructions properly. If you insist on using chemicals, *always* read everything on the label (even it's in such tiny print, you need a magnifying glass).

CHEMICAL PESTICIDES: KNOW WHAT YOU'RE USING

The following are some of the chemical pesticides sold in garden centers. They may be mixed together in various formulations, or combined with other products. Avoid their use wherever possible.

Insecticides

Acephate
POPULAR BRAND NAMES: Orthene, Ortran, Acephate 75 Turf
SIGNAL WORD: CAUTION

This kills many insects, but it is also moderately toxic to humans and wildlife. Residues stay in vegetables for two weeks, so do not use before harvest.

Chlorpyrifos
POPULAR BRAND NAMES: Dragon, Dursban, Ortho clor, Raid Home Insect Killer
SIGNAL WORD: WARNING or CAUTION

This is a source of organophosphate poisoning in fish and birds and is mildly toxic to humans, causing skin and eye irritation. Keep away from the water supply and ponds.

Dimethoate
POPULAR BRAND NAME: Cygon products
SIGNAL WORD: WARNING

This is highly toxic to birds and honey bees. Fish are moderately affected. It is mildly toxic to humans and other mammals as well.

Diazinon
POPULAR BRAND NAMES: Bonide Diazinon Soil Granules, Knox Out, many other products
SIGNAL WORD: WARNING (on spray cans); CAUTION (on granular forms)
Toxicity varies with formulation.

This is a commonly available insecticide, but one of the most controversial on the market. Nearly half the wildlife poisonings precipitated by home and garden use are said to involve diazinon. When applied to a golf course in Long Island, New York, in the 1980s, diazinon killed 700 geese and many other birds. As a result, the Environmental Protection Agency has banned its use on golf courses and turf farms. But it's still available to homeowners!

Horticultural Oils
ALSO KNOWN AS: dormant oil, dormant spray, miscible oils, horticultural spray oils
POPULAR BRAND NAMES: Many
CAUTION on some products.

Although horticultural oils are often considered "organic," some products contain chemicals. They suffocate pests such as spider mites and mealybugs on houseplants and when applied in early spring out of doors, before leaves appear on trees, shrubs, and flowers, they control aphids, scale, tent caterpillars, and other pests. These oils are toxic to fish and harmful if swallowed. They are also flammable. Do not use near heat or flame.

Malathion
POPULAR BRAND NAMES: Home Orchard Spray, Rose Spray, No Roach Spray, Bonide's Malathion Insect Control Spray, Malacide, Green Light Malathion
SIGNAL WORD: CAUTION

It can be harmful if swallowed or if it comes into contact with skin. Exposure to very high doses causes everything from unconsciousness to abdominal cramps in mammals, but low doses appear to have little effect. It is moderately toxic to birds and devastating for bees. Effects on fish vary with the species.

Methoxyclor
POPULAR BAND NAMES: Marlate, Methoxyclor 25, Dragon's Methoxyclor
SIGNAL WORD: CAUTION

This was developed as a substitute for DDT and is one of the only truly effective remedies for Japanese beetles. Although studies indicate that it has fairly low toxicity to animals and humans, one study in Spain links it to breast cancer.

Sevin
POPULAR BRAND NAMES: Sevin, Sevin Bug Killer Dust, Later's Sevin Liquid, various products
SIGNAL WORD: CAUTION or WARNING

Sevin is the common name for carbaryl, an ingredient in many pesticide products. It is toxic to honeybees and beneficial insects such as lady bugs so use sparingly and keep away from animals. Do not store it inside the home.

Herbicides

Dicamba
POPULAR BRAND NAMES: Banfel, Banvel, Dicazin, Fallowmaster, Mediben, Tracker, Trooper
SIGNAL WORD: WARNING

A herbicide that kills weeds before they sprout, this is not toxic to birds and bees and has little effect on fish. However, it is irritating to skin and eyes in mammals. One drawback is that its toxic effects stay at least two months in the soil. So it may kill anything planted after weeds have been eliminated—and have long-term residual effects.

Glyphosate

POPULAR BRAND NAMES: Roundup, Glifonox, Glycel, Kleen Up, Rodeo, Rondo, Vision

SIGNAL WORD: WARNING or CAUTION, depending upon the formulation

Glyphosate is considered one of the safest herbicides on the market because it is believed to break down quickly after application and not remain in the soil. However, its residual effect—and toxicity—is now being challenged by some critics. Glyphosate will kill most plants it touches for up to four weeks, so be careful where you spray it. Keep away from pets.

Mecoprop

POPULAR BRAND NAMES: Target, Sword, Chipco Turf Herbicide and others.

SIGNAL WORD: WARNING or CAUTION, depending on the product.

This is a general use herbicide that's often combined with other chemicals in a single pesticide product. It's irritating to skin and eyes in mammals, but not toxic to birds or fish. It's also linked to cancer in some studies, although this has not been statistically proven. One drawback is that Mecoprop takes two months or more to break down in the soil.

Triclopyr

POPULAR NAMES: Brush-B-Gon, Poison Ivy and Oak Killer, Garlon, Grazon, Pathfinder, Redeem, Turflon

SIGNAL WORD: CAUTION

This is one of the few products that controls poison ivy, and it breaks down quickly in the soil. But it's pretty toxic to fish, birds and some mammals. Avoid applying anywhere near lakes, streams, or ponds.

2-4-D

POPULAR BRAND NAME: Scott's Turf Builder Plus 2, Weed B Gon, other products

SIGNAL WORD: DANGER or CAUTION, depending upon the formulation

Widely used on lawns, this product used to be considered highly toxic to humans and pets. However, the United States Environmental Protection Agency recently decided that "there is no reason for concern about its short-term use." Even so, do not apply 2-4-D repeatedly.

Fungicides

Benomyl

POPULAR BRAND NAMES: Benlate, Benex, Tersan

SIGNAL WORD: CAUTION

A popular and relatively safe fungicide that treats black spot and powdery mildew on roses, phlox, and lilacs, it is toxic to fish and other aquatic creatures and will wipe out earthworms. Prolonged exposure is believed to cause eye abnormalities in children of horticultural workers.

Bordeaux Mixture
POPULAR BRAND NAMES: Kocide, Bordo-Mix, Tri-Basic, CPTS, Copper Bordeaux 22
SIGNAL WORD: WARNING or CAUTION

This is a mixture of copper sulfate and hydrated lime called "Bordeaux" because it's been used for years in French vineyards. It controls fungal and bacterial diseases like leaf spot, anthracnose, rusts, and molds, and is relatively safe when used as directed.

Captan
POPULAR BRAND NAMES: Orthocide, Ortho Home Orchard Spray, Ortho Garden Fungicide, Bonide Rose Spray, Bonide Captan, Green Up
SIGNAL WORD: DANGER or CAUTION, depending upon the formulation

Capstan is one of the most popular protectant fungicides around, which means it stops fungus diseases before they start, and is thus used extensively by seed companies. Captan is also sprayed on commercially grown apples. Its toxic effects are hotly debated. It does not harm birds or bees.

Ferbam
POPULAR BRAND NAMES: Fermate, Carbamate, Ferbam
SIGNAL WORD: CAUTION

This controls fungal diseases in flowers and fruit and is often combined with other fungicides, and has the same toxic properties as carbaryl.

Molluscicides

Snail and Slug Bait
POPULAR BRAND NAMES: Many
SIGNAL WORD: CAUTION or WARNING, depending upon the product and formulation

Baits containing metaldehyde or Mesurol are highly toxic to people, pets, fish, and wildlife, and should not be used. Those containing iron phosphate are less harmful.

ORGANIC PESTICIDES CAN BE TOXIC TOO

A pesticide is classified as "organic" if it's derived from natural sources and its residual toxicity does not linger in the soil, water, or atmosphere, but bear in mind that a label that says "natural" does not mean that product is completely harmless. Some organic pesticides are toxic to all kinds of living creatures and the environment. A few, made from poisonous plants, can be deadly. You may see these organic products on sale in garden centers:

Insecticides

Diatomaceous Earth

POPULAR BRAND NAME: Fossil shell flower

This is a white, abrasive powder that's made from ground-up fossilized shells of diatoms, which were once small sea creatures. It's normally dusted on plants (fill an old panty hose toe and hit the underside of leaves with it). DE works by puncturing holes in the shells of pests, causing them to dry up, and it's certainly effective at killing many kinds of insects and some caterpillars. Unfortunately, it also kills bees, so use sparingly. Preferably, dust this product on plants early in the morning before flowerheads have opened and bees are not as prevalent in the garden. It must be reapplied after rain (do not get the powder wet, as it clumps together) and always wear gloves because those little diatoms are razor-sharp. *Don't let pets nose around plants treated with DE.*

Hot Pepper Wax

POPULAR BRAND NAME: Hot Pepper Wax Insect Repellent, Bonide products

A mixture of cayenne pepper, repellent herbs, and food-grade paraffin wax, this liquid (which you mix with water and spray on) primarily repels rather than kills pests. It's effective against aphids, whiteflies, cabbage loopers, beet armyworm, and others. Right after spraying, plant leaves look cloudy, but the wax becomes transparent when dried. One application lasts about three weeks. It's also good for protecting plants in windy, hot conditions. It may repel deer, dogs, squirrels, and cats, and is preferable to using cayenne pepper powder, which may get into their eyes and cause blindness.

Insecticidal Soaps

POPULAR BRAND NAMES: Safer Insecticidal soap, various others
SIGNAL WORD: CAUTION

These contain salts of fatty acids, mixed with water and alcohol in various proportions. They are safe and easy to use on ornamental plants and vegetables, and they certainly do kill aphids, spider mites, mealybugs, some slugs, earwigs, scale (but only in their early stages), and whitefly. However, you have to keep reapplying these products and some are expensive. Many savvy gardeners find that a few squirts of a mild dishwashing liquid like Dove into a spray bottle of water will often do the job just as well as insecticidal soaps, but test on a plant leaf first because some products may cause leaf burn.

Neem Oil

POPULAR BRAND NAMES: Trilogy, Triact, Rose Defense, Green Light, Shield All, Bio-Neem
SIGNAL WORD: None. Not tested.

This is a relatively new product derived from the seeds of the tropical neem tree. Its chief virtue is an active ingredient called azadirachtin, which halts the life cycle of pests, so they eventually die (although it may take a number of days for this to happen). Neem oil also suppresses insects' appetites, so they stop feeding on plants. Effective on aphids, cutworms, fungus gnats, leaf miners, and thrips, it appears to have no effect on beneficial insects, such as ladybugs and spiders. It is best to buy cold-pressed neem oil, and mix it yourself with an emulsifying agent like a mild dish soap. (Look in stores that specialize in environmental products.) Ready-mixed sprays are less effective because the active ingredients dissipate quickly. Neem oil is also effective as a preventive fungicide.

Nicotine Sulfate

POPULAR BRAND NAME: Black Leaf 40
SIGNAL WORD: DANGER

This is one natural product that is extremely toxic to humans and other mammals. An extract of tobacco plants, it is usually sold as a liquid spray, or mixed with lime and turned into a powder. Nicotine sulfate is a nonselective pesticide, which means it will kill both good and bad bugs. It's particularly effective on aphids, whiteflies, leafhoppers, and thrips. Wear goggles and a mask, and use with extreme care. Because it is so harmful, most garden centers no longer sell it.

Pyrethrum

ALSO KNOWN AS: Pyrethrin
POPULAR BRAND NAME: Schultz Expert Gardener

Many gardeners assume that pyrethrum is a completely harmless pesticide because it's made

products, Ortho products, Bonide products, Spectracide, Miracle Gro Bug Spray
SIGNAL WORD: CAUTION

with oil extracted from the ground-up dried flowers of the African daisy *Dendranthema cinerariaefolium*, a popular container plant. However, pyrethrum may irritate skin, and some people and pets are allergic to it. Always wear gloves and avoid getting it into water you're going to drink. (Don't use the spray near a picnic table, for instance.) It works as a contact or stomach poison on many flying insects, including aphids, whiteflies, and houseflies, and won't harm most ornamental plants and vegetables. The pyrethrum products that contain piperonyl butoxide as well are more effective at killing insects.

Rotenone
POPULAR NAMES: Rotenone, Derris, DX, Rotacide, Bonide products
SIGNAL WORD: CAUTION or DANGER

This is another plant-based organic pesticide that is assumed to be 100 percent safe, but in fact isn't. Rotenone is derived from derris root (a kind of pea) and generally sold as a powder. It's also sometimes mixed with pyrethrum. It kills beetles, caterpillars, other chewing insects, and also fleas and ticks (so is formulated as a spray to use on pets too). While rotenone washes away quickly from the soil and leaves no harmful residues on food or vegetables, it is very toxic to fish. Don't use near ponds. Always wear gloves, avoid inhaling it, and buy it in small amounts because its effectiveness declines over time.

Ryania
POPULAR BRAND NAMES: Ryania, Natur Gro R-50, Natur Gro Triple Plus, Ryanicide
SIGNAL WORD: CAUTION

Derived from the ground stems of a tropical American tree called *Ryania speciosa*, this is primarily useful on vegetables and fruit trees. It kills fruit moths, coddling moths, corn earworm, European corn borer, and citrus thrips. Although classified as a botanical insecticide, it is quite toxic to birds and some mammals. Its effect on humans has yet to be studied.

Sabadilla
POPULAR BRAND NAMES: Red Devil, Natural Guard
SIGNAL WORD: None. Not tested.

Derived from the seeds of tropical sabadilla lily *Schoenocaulon officinale*, this is probably the least toxic of all botanical insecticides. It kills caterpillars, leafhoppers, thrips, stinkbugs, and squash bugs and leaves no residue behind on leaves. Avoid inhaling sabadilla, however, as it causes sneezing and eye irritation. It's also highly toxic to bees.

Tobacco Dust
POPULAR BRAND NAME: Bonide Tobacco Dust
SIGNAL WORD: CAUTION

Tobacco dust is a milder version of nicotine sulfate that's aimed at home gardeners. Sold as a powder in bags, it normally contains 5 percent nicotine while the rest is filler. It kills leafhoppers, thrips, and some other soft-bodied insects, but will also knock off beneficial bugs. Roses, commonly affected by thrips, may turn black after being dusted with this powder, so it's a good idea to rinse off their leaves.

Herbicides

Corn Gluten
BRAND NAMES: Bio weed, Corn Weed Blocker, Supressa, Safe 'n Simple, Turfmaize
SIGNAL WORD: None

Corn gluten is a hot new product that's gaining interest because it kills weeds, is 100 percent natural, and is completely nontoxic. A by-product of cornstarch, it's usually sold as a fine yellow powder (or in pellets) in 5-lb or 25-lb (2-kg or 11-kg) bags. Sometimes corn gluten may also be mixed with soil additives such as bonemeal and potassium sulfate. Its primary use is on lawns, but it can be applied to stop weeds elsewhere in the garden too. Because it's new, corn gluten can be hard to find at garden centers. Look in the weed and feed section. (See page 90 for instructions on its use.)

Horticultural Vinegar
POPULAR BRAND NAMES: Scotts Eco Sense Weed Control Spray
SIGNAL WORD: WARNING

Horticultural vinegar (acetic acid) is a brand-new weapon against weeds. It's increasingly used by municipalities that have banned chemical pesticides. It works best as a spot herbicide on broadleaf weeds like dandelions. While acetic acid is considered nontoxic, it can cause eye irritation.

Moss and Algae Killer
POPULAR BRAND NAMES: Safer Moss and Algae Killer and others
SIGNAL WORD: WARNING

This stuff, usually sold as a spray, is similar to insecticidal soap, but is formulated to kill algae, lichens, and mosses in lawns and on decks or patios. Don't use it on plant foliage to kill insects. It is entirely nontoxic, but is more effective on hard surfaces than on lawns. It tends to be expensive.

Weed and Grass Killer
POPULAR BRAND NAME: Safer Sharp Shooter
SIGNAL WORD: CAUTION

Another product that's made with biodegradable fatty acids and formulated to kill weeds and unwanted grass, this works best when applied as a spot herbicide on shallow-rooted, young plants. It will not kill older established plants with very long roots (like dandelions) and must be applied under specific conditions. Be sure to read the label.

Fungicides

Lime-Sulfur Solution
POPULAR BRAND NAME: Kaligreen and others
SIGNAL WORD: DANGER

A combination of hydrated lime and sulfur, and sold as a liquid, it is best applied as a dormant spray early in the spring when buds are swelling but not yet open. It controls powdery mildew on ornamental flowers and fruit, but also can be effective against scale, provided it's applied before the insects have matured and developed hard, horny shells. Always wear a mask and gloves as this stuff is highly caustic to skin, and don't use when temperatures are above 75°F (24°C).

Neem Oil
POPULAR BRAND NAME: Neem, Greeneem

Made from the tropical neem tree, this is useful mainly in preventing powdery mildew and other fungal diseases before they start, and is less effective once diseases get established. See page 128.

Sulfur

ALSO KNOWN AS: Wettable sulfur, wettable dusting sulfur, garden sulfur
POPULAR BRAND NAMES: Safer Garden Fungicide, Flotox, Garden Sulfur, Magic Sulfur Dust
SIGNAL WORD: CAUTION (in some products)

A fine, yellow, inexpensive powder made from ground-up sulfur rock, which is very useful to gardeners. It's been around for years and is relatively harmless, although you should avoid touching it. You can acidify soil with sulfur and dust it on flowers and vegetables to control powdery mildew, black spot, rust, and other diseases. At the end of the summer, put a bit of sulfur into a brown paper bag and toss dahlia, gladioli, canna tubers in it before storing them in a cool basement for the winter. The sulfur stops such plants from turning moldy. Micronized sulfur has the smallest particles and adheres better to plants, but it is more expensive.

Biological Controls

These aren't pesticides. They are bacteria that work by increasing disease organisms that occur naturally in the soil and on plants. Quite simply, pests ingest the disease we have stimulated artificially and they die. Biological controls are usually easy to use, they cause no residual damage, and humans, animals, and birds can't catch these diseases themselves.

Bacillus popillae

ALSO KNOWN AS: Milky spore disease, milky spore powder, BP
POPULAR BRAND NAMES: Doom, Japidemic, Grub Attack, Safer Grub Killer
SIGNAL WORD: None

Bacillus popillae is a powder of spores that is applied to lawns in spring to control grubs of Japanese beetles and June bugs. (Raccoons and skunks love these grubs and tear up lawns looking for them.) Grubs gobble the powder, get infected, and die. It may take up to three years for milky spore disease to spread through your soil, and while it kills grubs, it doesn't prevent them. If the beetles lay more eggs, there'll be more grubs. Spores become inactive easily and the container should be stored in a cool place. If you suspect it's been sitting around on a garden center shelf, don't buy it.

Bacillus thuringiensis

ALSO KNOWN AS: BT
POPULAR BRAND NAMES: Dipel, Thuricide, Bactur,

Usually sold as a dust, BT has been around for years and is widely used because it's harmless to humans,

Mosquito Attack, Safer Caterpillar Killer, Bacthane, Biotrol, Vectobac
SIGNAL WORD: CAUTION

animals, food crops, and useful insects. It kills many kinds of caterpillars and the Colorado potato beetle by giving them a bad case of indigestion. Most useful on vegetables, whose leaves must be thoroughly covered (top and undersides) with BT, it can be applied right up until the day of harvest. This product is often sold mixed with other products, such as insecticidal soaps.

Organic Physical Controls

Insect Traps
POPULAR BRAND NAMES: Sticky Strips, Tanglefoot, Flying Insect Trap

Updated versions of flypapers, these are usually flat pieces of yellow card coated with a sticky substance and sold with wire holders. They trap white flies, aphids, houseflies, fungus gnats, fruit flies, leafhoppers, and houseflies (different formulations are used for different bugs) and are always yellow because that color is believed to lure most insects. These are particularly useful on houseplants. Put a sticky strip in a clear plastic bag together with a plant that's infected, seal the top, and leave for a couple of weeks. These traps are less successful outdoors because on windy days, garden detritus and dust, instead of the insects, are blown onto the strips. Position them in sheltered areas, right next to affected plants.

Pheromone Traps
ALSO KNOWN AS: Sex traps, Japanese beetle traps, pheromone lures
POPULAR BRAND NAMES: Bag-a-Bug, Japanese Beetle Attack

A container (usually plastic or cardboard) with a large, bug-attracting yellow strip on top and impregnated with the female Japanese beetle sex pheromone, it lures several kinds of insects that fall into the container and die. But the problem with this device is that it may attract too many bugs to the garden, and the phereomone can't handle them all at once, so they fly around, looking for things to land on. Position this trap well away from any plant that you don't want to be eaten.

Toad Houses

It's too bad so many people find toads creepy because they probably get rid of more garden pests than anything else. They're most active at night, dining on slugs, cutworms, sowbugs, caterpillars, and various beetles. To encourage toads to move in (you can't simply go out, get yourself a toad, and plonk it in the garden), create cool, moist, shady hiding places. They love upturned broken clay flowerpots, tucked under big-leafed plants. A couple of rocks propped up in an upside-down V makes a good home for them too, and many potters sell unobtrusive, handmade toadhouses that look delightful—and usually delight Mr. and Mrs. Toad too. Toads need a water source close by. Be careful of skewering toads with a garden fork or a trowel while working in the soil around plants, and try not to disturb them. They will crawl out, blink at you, and look thoroughly baffled. Cats and dogs are curious about toads, but usually don't try to play with them, seeming to know that some kinds of toads can be poisonous. Frogs, while useful around ponds and fun to hear croaking on hot nights in summertime, don't eat nearly as many insects as toads.

COMMON GARDEN PESTS AND HOW TO COMBAT THEM

There are an overwhelming 2,000 insects classified as "pests" in North America by horticulturists, agronomists, and scientists. They may make a nuisance of themselves at any stage of their development as larvae, immature beetles, leafhoppers, or flying or crawling insects. Here are some of the most common offenders.

Ants

Ants can be alarming because they toss up piles of soil in flower beds and between cracks in pathways. They also fly around in swarms, and some species sting or bite. However, they don't actually attack plants. When a plant wilts near an ant heap, it's because their tunneling has caused gaps in the soil, leaving roots dangling and dried out. And if ants are climbing higher on plants, it's usually because something has attracted them there. Often it's aphids. Ants don't eat aphids, but they like dining on their honeydew, and they will actually protect the aphids to ensure a steady supply. Ants are also attracted to peonies because as the flower heads open, they exude a nectar.

Remedy
Organic
- Smash and destroy a big ant heap with a garden fork or spade, but be care-

ful of stinging, angry ants flying out as you do this. Dig thoroughly, get rid of their tunnels, and flatten the area. The ants will usually move on to more hospitable pastures. When they're a problem between cracks in pathways, pour a kettle of boiling water down the hole.

Chemical
- Chlorpyrifos

Aphids

One big problem with aphids is that you usually can't see them. They're microscopic sucking insects, teardrop-shaped, and they may have wings or not, depending on the stage of their development. Some also act like chameleons, changing color (which may be yellow, pink, black, or green) to match the host plant. Whatever their color, aphids are huge nuisances, sucking the lifeblood out of many different plants. They also carry viruses that infect many plants (including tulips), and their honeydew attracts ants, so they are best eliminated.

Telltale signs are leaves that are puckered, curled under, or stunted; the entire plant wilts and/or turns pale. Tender growing tips of leaves and buds are scrunched together, sticky with honeydew, and unable to open. There may also be a sooty mold.

Remedies

Organic
- Train a jet of water from the garden hose on them.
- Spray with soapy water (or insecticidal soap) three times, at three-day intervals. Pay the most attention to undersides of leaves.
- Encourage ladybugs, which eat thousands of aphids.
- A mulch of aluminum foil, laid around the base of plants, is said to deter aphids because the light shining off the foil confuses them, so they go elsewhere.

Semiorganic
- Mix a plastic sprayer full of pyrethrum, with 1 tbsp (15 mL) of isopropryl alcohol. Spray on plants.

Chemical
- Acephate, malathion

Borers

The larvae of several kinds of moths and beetles will bore into the stems of plants, shrubs, and trees. They look like white, gray, or pink caterpillars with brown heads. A telltale sign, in soft-stemmed plants, is a stem that suddenly starts wilting, or it may keel right over and actually break. In tougher, thicker branches, you'll notice little holes (and sometimes a pile of sawdust) somewhere along the branch. Borers weaken plants and invite disease. They often affect bearded irises and squash vines.

Remedy
Organic
- If you suspect borers, cut a stem open lengthwise with a Swiss army knife. You'll probably find the culprit lurking inside. Cut down all stalks of the plant at the base and burn them. (This is particularly important for bearded irises, as borers will get into the tuberous roots and ruin them.) In the case of squash, remove the entire infected stem, but you can leave the rest of the plant as is. On large branches of shrubs like lilacs, poking an opened coat hanger through the borer hole and pushing it around, upwards, and downwards may kill the little creeps. If not, remove the affected branch. Lilacs sometimes fight off attacks by borers on their own, but often the only solution is to remove the entire plant or shrub. If bearded iris keep getting attacked by borers, plant Siberian iris instead.

Chemical
- Spraying affected areas with horticultural oil may block borer holes and suffocate them.

Caterpillars

Everyone has fallen in love with butterflies. But one fact that's overlooked in our eagerness to create butterfly gardens is that butterflies and moths start out life as caterpillars and that at this larval stage, they love dining on juicy green leaves. Learn to identify the caterpillars that metamorphose into beautiful butterflies in your garden and dispatch the dullards. But beware that both "good" and "bad" butterflies may nibble on your plants. The welcome ones can in fact be voracious eaters when still at the caterpillar stage, and you may have to limit their numbers or move them elsewhere. Yellow-and-black swallowtail butterflies, for instance, are glorious to have around, but as caterpillars (when they wear an equally striking uniform of black, yellow, and pale green stripes) they are often called parsley worms and they love dining on herbs and vegetables, particularly parsley, cilantro, dill, and carrots. (They're also partial to a wild member of the carrot family, *Daucus carota*, commonly known Queen Anne's lace. So if you have some of this common weed growing in your garden—and can't bear the thought of banishing the beautiful swallowtail butterflies—you might want to move the offending caterpillars there, away from your herbs.)

Caterpillars of the most idolized butterfly of all, the monarch, dine mostly on wild milkweed, *Asclepias tuberosa*, and some other flowers, but are not usually a problem in gardens. Signs of caterpillar damage may be chewed or skeletonized foliage or leaves rolled up and fastened into a little bundle with webbing.

Remedy
Organic
- Handpick and squish the caterpillars, or drown them in soapy water. In bad infestations, cover the leaves, especially the undersides, with Bt. *Bacillus Thuringiensis*. You can also dust the foliage with an old pantyhose toe filled with diatomaceous earth.

Chemical
- Acephate, carbaryl (Sevin)

Monarch butterflies may land on zinnias (above) and some other flowers but they particularly like common milkweed *Asclepias tuberosa*.

Cutworms

Fat, creamy-coloured grubs that are the larvae of a rather handsome night-flying moth. Mostly a nuisance around small tomato plants, cutworms chew right through stems at night, felling the entire plant. Annual flower seedlings may also be guillotined.

Remedy
Organic
- Dig around in the soil before planting. The grubs are easy to locate (they curl up when exposed to light) and squish. When putting in tomato seedlings, push a matchstick or small twig in the ground right next to the plant stem, so that they touch. The cutworm won't want to chew through both, and will move elsewhere. (This is vastly easier and more effective than cutting the bottoms out of paper cups and positioning the cups around tomato stems, as is advocated in many gardening books.)

A twig placed right next to a tomato stem will prevent cutworms from demolishing the plant.

Chemical
- Chlorpyrifos, Diazinon, Sevin. Do not use in vegetable gardens. It is seldom necessary to treat soil with chemicals to eradicate cutworms, however, since infestations aren't usually severe. And when they get past the cutworm stage and turn into moths, they burrow out of the soil and fly away anyway.

Earwigs

One of the biggest garden pests, particularly if you put down lots of mulch around plants, because they love damp, dark hiding places. They look like long, hard-shelled brown beetles, with nasty pincers on their rear ends, and they can actually bite, although horror stories about them are greatly exaggerated. A few earwigs can be beneficial in a garden because they munch on another annoying intruder, the aphid. But if great mobs of them are chewing everything in sight, take action. The best time to check for earwigs is at night, with a flashlight. They're often rushing all over the place in astonishing numbers. They are much worse during rainy, cool summers.

Remedies
Organic
- Try 6-in (15-cm) lengths of old garden hose, placed randomly under plants. Then in the morning, dump these into a pail of soapy water. You can also put scrunched-up balls of damp newspapers around, then dunk these into the pail. (Wear gloves as the newsprint comes off.) Get some shallow, empty containers with steep sides, such as tins or margarine tubs. Pour in 1 tbsp (15 mL) each of vegetable oil, soy sauce, and molasses. Mix. Place containers near plants that earwigs dine on. Bend a leaf or two over the containers to stimulate earwiggy interest, but be sure that once they topple in, they can't climb out again. And if earwigs keep coming back, remove the mulch and grow things in bare soil.

Chemical
- Chlorpyrifos, Diazinon, Sevin

Flea Beetles

These are tiny and black, they jump around a lot like their namesakes, and they make pinholes in leaves of plants. This stippling effect looks unsightly, but actually doesn't cause a lot of damage. As the summer wears on, flea beetles usually disappear.

Remedies

Organic
- Spray with neem oil. Flea beetles are worse when it's dry, so try spraying plants with the garden hose in the middle of the day. Growing tansy is said to deter flea beetles, but don't do this in a small garden because the plant is very invasive.

Chemical
- Diazinon

Grasshoppers

These big, brownish insects with long legs chew large holes in leaves. You can also hear them jumping about, particularly in long grass. Cats love grasshoppers—they're fun to catch—but they are a huge problem in parts of the South and will also find their way into northern gardens during droughts.

Organic
- None; pray for rain because grasshoppers hate wet weather.

Chemical
- Acephate, as soon as you see them, but only in serious infestations.

Japanese Beetles

Shiny, hard-shelled beetles, usually a brilliant blue-green. They often suddenly appear in alarming numbers on rose leaves and flowers (also other plants), chewing and making a mess with their feces. The grubs of these beetles are just as troublesome because they hatch in lawns, and raccoons and skunks love to tear out turf, looking for them to munch on.

Remedy

Organic
- Pick off beetles in early morning and drown them in soapy water. To combat grubs, apply *Bacillus popillae* (also known as milky spore disease and sold under the brand name Doom or Safer Grub Killer) to the lawn. This will kill the grubs but is not a preventive treatment. It works only if bugs are present. *Bacillus popillae* must also be stored in a cool, dark place

like the fridge because the spores quickly become inactive in warm temperatures. Don't buy it if it's been sitting for ages on the garden center shelf. In some areas, Japanese beetles are developing a resistance to milky spore disease and you may have to try beneficial nematodes.

Chemical
- Diazinon, Methoxyclor, Sevin

Leafhoppers

There are many kinds of leafhoppers, but they're all recognizable by the way they carry their wings on top of their bodies, like a roof. Mostly small (although the grasshopper is a big cousin), they are often colorful and move suddenly, often sideways. They suck juices from plants and stems, and excrete copious amounts of honeydew. In large infestations, plant leaves may look glazed and shiny.

Organic
- Spray with insecticidal soap mixed with rubbing alcohol. Mix 1 tbsp (15 mL) of alcohol into each gallon/liter of soapy solution. Spray three times, every three days.

Chemical
- Diazinon, Sevin

Leaf Miners

White squiggly lines on leaves are a sign that leaf miners have invaded—so are raised yellowish blisters—and, unfortunately, by the time you spot them, it's too late to do anything. These troublemakers lay bundles of white eggs on the undersides of leaves and it's necessary to remove the clusters in very early spring to stop the little worms from tunneling inside. However, apart from looking unsightly, they don't do much damage to many plants. The leaves of columbines *Aquilegia* are particularly prone to attack, and some gardeners, unable to eradicate leaf miners from these spring flowers, learn to live with them. ("I now regard my leaf miners as abstract paintings on the columbine's leaves," says one.) But when leaf miners attack hollies *Ilex*, lilacs, or chrysanthemums *Dendrathema*, they may weaken plants and slow down their growth. Some vegetables are also affected.

Remedy

Organic
- Don't leave garden detritus lying around in fall because leaf miners overwinter there. Spray the plants with neem oil in early spring.

Chemical
- Malathion

June Bugs (or Beetles)

These big, brown, hard-shelled bugs are mostly a scary nuisance because they hatch on warm nights in May or June (depending upon where you live) and then behave like kamikaze pilots. Their larvae are the main problem because, like the larvae of Japanese beetles, these fat little grubs are a lip-smacking treat for raccoons and skunks. And because June bugs like laying their eggs in the compacted thatch lurking under lawns, when those grubs hatch out in spring, raccoons and skunks start rummaging around, looking for a meal and pulling up the grass.

Remedy

Organic
- Same as for Japanese beetle grubs.

Chemical
- Diazinon

Lily Leaf Beetle (*Lilioceris lilii*)

An introduced pest from Europe that's gaining hold in North American gardens, this is an elongated, shiny, bright red beetle with black antennae. It attacks primarily Asiatic lilies (daylilies are ignored), often eating every leaf. It may also devour foliage of lily of the valley *Convallaria*, potatoes, hostas, crown imperial *Fritillaria*, and other plants. It overwinters in soil and plant debris, and lays up to 300 eggs in spring; ten days later, larvae hatch and start munching.

Remedy

Organic
- Handpick the beetles and drown them in soapy water.

Chemical
- Carbaryl (Sevin) may work, but controlling this beetle is difficult because it has a hard, protective shell and no known predators.

Mealybugs

Small, sucking insects that are soft and oval-shaped, with telltale little "legs" encircling them. They're often covered in a cottony substance and generally located in the axis of leaves, stems, and branches. A huge problem on houseplants (usually because of insufficient humidity indoors), mealybugs affect plants outside, too, and may keep on producing a discouraging number of egg masses.

Remedy
Organic
- Try blasting mild infestations off with the garden hose, or dabbing the insects with a cotton swab dipped in alcohol. If the problem is serious, spray with insecticidal soap and improve airflow around plants. If mealybugs persist, remove and destroy affected plants.

Chemical
- Malathion; horticultural oil sprayed on plants may suffocate them.

Nematodes

There are good and bad nematodes. They're all microscopic worms and the baddies may produce hard lumps and swellings on plants and stunt their roots. If growth is poor, parts of a plant are dead or yellowing, or there are few flowers, the culprit could be these sucking, thread-like creepy-crawlies. Unfortunately, because nematodes are so unobtrusive, you can usually spot only the symptoms, not the worms themselves.

Remedy
- Dig up plants and check their roots. Bulbous and tuberous plants are often more likely to fall victim to nematodes than other plants. It's advisable to throw out anything affected by nematodes. If the problem persists year after year in a certain area, the only course of action may be to remove the affected soil and replace it, since chemical treatments are seldom effective. A humusy soil, rich in organic matter, will deter nematodes. And don't be too quick to blame plant problems on these little

tunneling creatures. Run through other possible causes first.

Scale

Scale isn't crud found inside the kettle. Nor is it a plant disease (as a lot of people presume.) It's actually is a horrid insect with a hard, horny casing that sucks sap from stems. If there's a sticky substance dripping under plants and they're wilting, the problem may be scale. Check stems and branches, particularly on the undersides. They'll look thickened with raised, fattened sections where scale is clinging on. Houseplants are often attacked. Outdoors, scale may get a grip on euonymus, geraniums, peonies, and cannas. The casings are surprisingly tough, so this pest can be difficult to eradicate.

Remedy
Organic
- On houseplants, you can try scraping off the limpet-like casings with the back of a knife or your fingernail, then spray the whole plant with soapy water. Another remedy is to dab the casings with a cotton swab dipped in rubbing alcohol. With larger, outdoor plants and shrubs, cut off infected stems and throw them out. You can also try dousing the entire plant with horticultural oil.

Slugs

Slugs are the numero uno problem for most gardeners, particularly if the weather is wet and you mulch a lot around plants (like earwigs, slugs love damp hiding places). Don't ignore slugs. They're voracious varmints, they multiply rapidly, and plants that they attack quickly show big ragged holes in their leaves, then can become completely skeletonized in a few days. Slugs usually aren't visible during the day, but their silvery trails leading along garden paths and under plants are often evident.

There are several kinds of slugs and the biggies (found on the West Coast) look revolting, but it's the small ones that do the most damage. They work at night, rasping holes in leaves with their tongues. Their favorite dinner is any green leafy plant (they hate hard-leafed plants like rhododendrons) and they're particularly partial to basil, nasturtiums, Swiss chard, lettuces, the cabbage family, and hostas. They also seem to be most prevalent just before rain, so that's the best time to go out and catch them in the act.

Remedies

Organic

- Handpicking: Go out at night with a flashlight. (Wear gloves. Slugs are horribly slimy and the guck is hard to wash off.) Drop them into a pail of salty water.
- A plastic sprayer with 1 tsp (5 mL) of ammonia added: Spray directly on slugs. Make sure they die (you may have to do this twice). They are remarkably tough.
- Saucers of beer: This tried-and-true remedy does indeed kill slugs (they drink themselves silly, fall in, and drown), but it's expensive. What's cheaper is mixing 2 tbsps (30 mL) of brewer's yeast (sold at health food stores) with 1 tsp (5 mL) of sugar into a 16 oz (500 mL) empty yogurt container filled with water. Pour this into saucers (or special slug traps, sold at garden centers, which are preferable because they have lids that keep out the rain) and sink into the soil, so the rim is at ground level. One problem with these concoctions is that they get smelly. Be sure to check them often.
- Copper tape: Slugs don't like crossing it. This is a good choice for plants in containers because you can wrap a length of it around the container. Copper tape is sold at some garden centers. (One brand is called SureFire Slug and Snail Copper Barrier.) You can also put strips of copper sheeting around the edge of flower beds, but this stuff is pricey, and when the tape gets dirty, the slugs will start crawling over it. And it won't deter slugs that are already within the ring of tape, lurking in the soil.
- Diatomaceous earth, coffee grounds, salt, wood ash, or crushed-up eggshells sprinkled around the base of plants.
- Removing lower leaves of plants, so slugs don't have any damp hiding places. Check regularly under rocks, pots, and in crevices between paving stones for them too.
- A brew made in fall of chopped-up leaves of wormwood *Artemisia absinthium*, poured over plants at that time, may kill any slugs lurking there.
- Hostas that are slug resistant: There are several varieties now. Among them: 'Abiqua Drinking Gourd', 'Invincible', 'Serendipidity', and 'Sum and Substance' (a beautiful, big variety that looks wonderful in a container. In fact, if you're persistently bothered by slugs, growing all hostas in containers is the way to go.)

Chemical

- There are many slug baits. Avoid those containing metaldehyde, which can cause kidney damage in kids and pets. Varieties containing iron

Slugs love hosta leaves. The best antidote is to pick slug-resistant varieties or to grow hostas in containers.

phosphate are less harmful. Also remember that birds dine on slugs, and will ingest the poison too.

Snails

Treat in the same way as slugs.

Sowbugs

These look like little gray armadillos, and they congregate, often in large numbers, in dark, damp places. They immediately curl up into tight balls when disturbed or exposed to light. Although many experts insist that sowbugs don't bother plants, they can be chewing pests, like earwigs. They often live in coarse gravel and cracks between paving stones, but their favorite garden habitat is beneath flowerpots. During the night, they will make their way up through the soil in that pot to dine on the plant. They love dahlias, basil, and anything else that's green and leafy.

Remedy
Organic

- Squash sowbugs if you're quick enough. (Some scurry off smartly.) Then scatter wood ash or diatomaceeous earth around areas where they hang out. However, the best antidote is prevention. To get rid of them under containers, raise the containers on legs or pieces of broken clay pots, so that air and light can pass underneath. The little nuisances should then move on.

Spider Mites

These tiny spiders are often invisible because they're so small. What you may notice is their webs, woven over and over on plants, so they resemble little cotton balls. There are many different kinds of mites, but all of them suck juices out of plants, causing speckled, sickly, pale foliage and spots that look corky and brown. They are often a problem on bulbous plants, such as tulips and gladioli, which may not bloom because the whole plant has become infected. Mites are worse when the weather's hot and dry. A dry spring, without much rain, triggers infestations. They also love attacking houseplants grown in too-dry homes, without the benefit of a humidifier, and the hotter it is, the more they will multiply.

Remedy
Organic
- It's important to act as soon as you notice a telltale stippling effect on plant leaves. Hold a sheet of white paper under a leaf and tap the top. Tiny moving specks on the paper means mites. Outdoors, train the hose on plants and give them a forceful blast, from a distance of only a few inches, paying attention to the undersides of leaves. Indoors, spray with insecticidal soap several times at three-day intervals, and try encasing the plant in a sealed plastic bag for a week or two after this treatment. Adding a few drops of seaweed extract to the soap solution is said to be beneficial. Don't crowd plants together, as mites leapfrog from leaf to leaf. They will also hitch rides on anything, even hands and tools. If plants are severely infested indoors, throw out the plants as mites can get into drapes, furniture, and bedding.

Chemical
- Horticultural oil sprays will suffocate mites.

Thrips

These tiny, darting black insects, common on flowers, are particularly attracted to pale-colored blooms and are fond of roses. Thrips will hop away quickly if disturbed, so they can be hard to identify. Shake a possibly infested flower over a sheet of paper without disturbing the plant too much. If you notice pinpricks of dark fecal matter mixed with white stuff (which are dried-up digested bits of plants), it's a sign of thrips. They discolor flowers, make flower buds turn brown and die (often without opening), and also feed on leaves, leaving tiny black spots of excrement on their undersides. The most telling sign of thrips is those little flecks of white on everything.

Remedy

Organic
- Thrips are difficult to control because they burrow deeply into petals and leaves. Spray with insecticidal soap as soon as you spot them and prune off all affected flowers and leaves. Thrips like dry conditions, so make sure plants are adequately watered. Sticky traps placed around plants early in the year can nip them in the bud. It's also important to clean up around plants in fall.

Semiorganic
- An application of pyrethrum will paralyze thrips, but it must actually touch them to be effective. Rotenone sometimes helps.

Chemical
- Acephate, applied in spring, in cases of chronic infestations.

Tomato Hornworms

If hornworms show up in your garden, it'll be a shock the first time it happens because they're huge—at least the length of a pinky finger, and often longer—with a scary (but harmless) horn-like protuberance sticking up from their rear ends. And at the other end, there's this weird face peering at you. Almost exclusively a pest on tomatoes, hornworms have greenish stripes, in two shades, with red or black on their tails, and they blend in so well with tomato plant leaves that the first sign of them may be a munching. noise. Sit quietly in the garden on a summer's day and you can actually hear these monster caterpillars tucking into their dinner. They are voracious eaters when mature, often denuding an entire tomato plant in a couple of days. However, one point in hornworms' favor is that they usually materialize late in the season, when tomatoes are maturing, and they prefer the leaves to the ripening fruit. Those mouthwatering, reddening tomatoes come under

> **ALERT!**
> *What honeydew is*
>
> In supermarkets, "honeydew" signifies something desirable: a tasty melon. But in the garden, the same word is used to denote nasty secretions from insects. These secretions may be sticky or liquid, colorless or orange. But if honeydew is present—clinging to either the top or underside of leaves, or to buds, or dripping underneath the plant—it's an indicator of a bug infestation.

attack only when hornworms have polished off all the leaves surrounding them. Hornworms are host bugs to trichogramma wasps, which are useful to have around in the garden because they feed on the eggs of many garden pests, such as cutworms and tent caterpillars. If you see a hornworm with small clusters of white eggs on its back, don't squish it, but get rid of its unadorned buddies.

Remedy

Organic

- Handpick large ones and drown them. Spray small ones with Bt *Bacillus thuringiensis* or neem oil. Look for tiny greenish balls on the undersides of tomato leaves in the spring. They're the eggs, and those leaves should be removed.

Chemical

- Sevin, but this is seldom necessary because these caterpillars are large, thus easy to spot and pick off.

Tomato hornworms are huge and fascinating to watch. But they can denude a tomato plant in days.

Whiteflies

These tiny, white insects suck juices from plants, affecting every part and stunting their growth. Because they tend to congregate on the underside of leaves, often the first time you notice whiteflies is when you brush against some infected plants, and a white cloud rises into the air. These insects are the bane of commercial greenhouse growers, and are one reason you should always buy bedding plants from a reputable nursery that makes sure its stock is whitefly free. Once whiteflies get into gardens, they are difficult to control and it's best to destroy heavily infested plants. They can also be a huge problem on houseplants.

Organic

- Yellow sticky traps early in the growing season. Later, mix 1 cup (250 mL) of rubbing alcohol and $1/2$ tbsp (7 mL) of insecticidal soap into a jug of water and spray on plants at weekly intervals.

Chemical

- Malathion

THE FOUR KINDS OF PLANT DISEASES

- *Cultural disorders* are caused by the growing environment and are the easiest to correct.
- *Fungal diseases* are contagious and tend to spread over the entire plant, then move on to other ones. They can worsen in humid weather but can sometimes be stopped in their tracks with treatments or improved gardening practices.
- *Bacterial diseases* occur when microscopic organisms invade the plant. There is no cure for bacterial infection. Affected plants should be removed and destroyed.
- *Viral diseases* are parasites that get inside plants and multiply. They can lie dormant for years within plants (even dead ones), then be spread to other plants by host insects, such as aphids. There is no cure for viruses. Afflicted plants should be removed and destroyed.

What Chlorosis Is and What It Means

Chlorosis is the yellowing of leaves, and it's the most glaring symptom of cultural disorders or fungal diseases. However, it's important to notice where and how the yellowing is affecting the plant.

- *If the entire plant is going yellow,* there could be a lack of nutrients in the soil, or high temperatures, or the light is too strong in that location.

- *If only the bottom leaves are going yellow* and the plant is wilting, it may have root rot.

- *If only the small, new leaves appear yellow,* your plant needs manganese or iron, or there's not enough light where it's growing.

- *If older leaves keep going yellow,* the soil lacks nitrogen or potassium, or it's too clayey and compacted.

- *If only the edges of leaves turn yellow,* the plant needs magnesium and potassium.

- *If there's yellowing between the leaf veins,* the plant may need iron or manganese, or there's a lot of sulfur dioxide in the air.

- *If there are yellow, irregular spots,* it could be too much cold water hitting the plant or a fungal disease.

- *If the yellowing is in a mosaic (i.e., broken) pattern*, it could be too much cold water.

- *If leaves are turning yellow and drooping*, the plant is not getting enough water.

Other Noticeable Signs of Plant Disorders

- *If leaves look dead and brown at the tips or around the edges*, it may indicate a potassium deficiency, too much boron, too much fluoridated water, or the weather is too cold or too hot, or the plant has become too dry.

- *If the plant's leaves have brown spots or sections turning brown*, it may be getting too much cold water or have a fungal disease.

- *If edges or inner sections are looking brown and rotted*, the location may be too cold, the plant is getting too much cold water or too much light, or it has a fungal disease.

- *If leaves look pinched and curled under, and the plant isn't growing much*, the growing location is too cold, although this may also be an aphid problem. (See page 134.)

- *If leaves have spots that look rusty and/or moldy*, the plant has a fungal disease.

- *If the plant looks waterlogged and stems are mushy*, the growing location is too cold or too hot, or the water is too cold, or you've been watering too much, or it has a fungal disease.

- *If a seedling suddenly withers or collapses*, it has contracted damping-off disease and should be destroyed immediately.

When the problem is a nutritional deficiency, one quick fix (particularly for houseplants) is to use a foliar spray. In the garden, it's a good idea to check the pH (see page 43), then rectify the problem with organic matter added to the soil and/or fertilizers. Remember, too, that unseasonable weather patterns are a frequent cause of cultural disorders. If it's too cold or wet in spring, or too hot and dry in the height of the growing season, it will inevitably affect how well your plants grow.

The weather is also a precipitating factor in fungal diseases, which always make their presence felt in humid conditions, either when it's cold and wet or hot and steamy. These kinds of diseases are also more difficult to correct than cultural disorders.

COMMON DISEASES THAT AFFECT PLANTS

Bacterial Wilt

This is a frustrating disease because plants may wilt, then recover, then wilt again. Finally leaves start turning yellow and the whole plant dies. Susceptible plants include dahlias, delphiniums (particularly so), nasturtiums, petunias, and zinnias. To confirm bacterial wilt, cut a stem open. If slimy stuff oozes out, remove the plant immediately and destroy it. There is no cure for bacterial wilt and it can spread easily, so wash hands well after getting rid of the victims.

Black Spot

Gardeners who love roses are the most familiar with this tiresome problem. Small black spots appear on rose leaves, surrounded by yellow or orange rings. Leaves then turn yellowish pink and drop off. The disease is precipitated by humid weather, lack of air circulation in the garden, insufficient sunshine, and growing too many of one type of rose in the same place. But even in perfect conditions, some roses (particularly hybrid teas) will develop those darn spots anyway. When buying roses, look for disease-resistant kinds. There are many available now, and this information is usually included on plant tags.

Remedy
Organic
- Remove affected leaves immediately, then spray with baking soda (sodium bicarbonate) mixed in water at a ratio of about 1 tbsp (15 mL) of soda to 1 gal (3.75 L) of water. (Some people add 1 tbsp/15 mL of vegetable oil, too, to help the concoction cling to leaves.) This treatment has been proven to be very successful at stopping black spot, but you have to keep doing it every week. Spray in the early morning, and repeat if it rains.

  Alternatively, spray with insecticidal soap or a wettable or liquid spray of sulfur, then keep making regular applications throughout the season. Some people find neem oil helps.

  Don't subject roses to blasts of cold water from the hose. Water them in the early morning and direct the hose nozzle around the base of the plant. They like a good soak once a week. Avoid getting cold water on their leaves. A mulch around the base can stop water splashing up. Keep roses out of the range of lawn sprinklers.

Chemical
- Benomyl, Captan

Blight

There are several kinds of blight. Some hit plants early, others late, but sudden wilting or browning, particularly in humid conditions, is usually a symptom. Many flowering plants can be affected, including dahlias, geraniums, peonies, iris, tulips, roses (rosebuds that fail to open, then turn brown and crinkly are a telltale sign), and zinnias. Blight also hits vegetables if it's cold and wet in spring or if there isn't enough dry sunny weather when they're maturing. (Squash vines may develop dark, smelly blotches on their leaves and the fruits will rot.)

Remedy

Organic
- This is difficult. Try spraying plants with manure tea or neem oil at the first sign of blight. Insecticidal soap or sulfur- or copper-based fungicides may help. Prevent problems from occurring by cutting off dead blooms promptly (blight hangs around in old blooms) and cleaning up dead foliage in fall. Water in the early morning, so leaves can dry off and aren't sitting around wet.

Chemical
- Benomyl, Captan

Fusarium Wilt

Commonly known as "damping-off disease," this affliction is dreaded by anyone who raises plants from seed. A fungus develops at the base of seedlings so that their stems shrink inward and may turn black. The fledgling plants then wilt and collapse. Vegetable plants, particularly tomatoes, are often affected. A whole tray of seedlings can be wiped out in a matter of hours once this disease gets a grip.

Remedy
- There is no cure for fusarium wilt. The moment you notice seedlings keeling over, pull them out. Even then, it may be too late to save the others. Prevention is the best cure. Use sterile seed mix, don't overcrowd seedlings or overwater them, and train a fan on the growing area to keep air circulating. Verticillium wilt is a similar disease. Look for seeds that are disease resistant. (They're marked "vf.")

Gray Mold (Botrytis Blight)

This can affect just about any plant during humid weather. The most common symptom is a fuzzy grayish mold that hovers above brown, spongy, rotten-looking patches on leaves and flowers. The mold may also make flower heads droop. Botrytis blight usually doesn't kill plants, but it looks unsightly.

Remedy
Organic
- Try spraying with manure tea or neem oil. The most important antidote, however, is to provide good air circulation. This disease only develops when plants are packed too closely together. Remove affected stems and destroy them.

Chemical
- Benomyl, Bordeaux mixture, Captan

Phytophthora Blight

This is a bacterial disease that particularly affects rhododendrons, causing their leaves to discolor and stems to die. To restrain this fungus, provide rhodos with well-drained soil. Working in some shredded bark seems to help. Avoid wetting leaves when you water. There is no cure or treatment.

Powdery (and Downy) Mildew

These two fungi look very similar and they're probably the most common fungal diseases in gardens. They produce a white or grayish powdery coating on leaves. (Downy mildew attacks in cool, wet weather. Its powdery partner hits late in the summer if it's hot and humid during the day but cold at night.) Either way, plant leaves, buds, and flowers are coated in whitish stuff or develop blotchy spots, and the entire plant may become so discouraged

Garden phlox (rear) is prone to mildew. Look for disease-resistant varieties since treatments seldom work for long.

and weakened that it won't flower or set fruit. Plants that are particularly prone to mildew include phlox, bee balm *Monarda didyma*, columbine *Aquilegia*, roses, zucchini, and squash. If you put houseplants outside in a humid summer, they may also contract mildew because they've been used to hot, dry central heating indoors.

Remedy

Organic

- The baking soda treatment (see Black Spot on page 151). Don't overcrowd plants or grow them in damp shade. Avoid splashing cold water on their leaves. Some people swear by a garlic spray. Position phlox and bee balm behind shorter plants, so that their moldy stems and lower leaves are hidden. But the best antidote is prevention: pick disease-resistant varieties.

Rust

A perennial problem on hollyhocks, rust can also affect many other annual and perennial flowers, including roses, snapdragons, black-eyed Susans, chrysanthemums, and sunflowers, and vegetables such as asparagus, beans, beets, and Swiss chard. This disease is easy to identify because leaves develop a rusty coating or stippling (which is often a startling orangey-red), sometimes accompanied by lesions. Then the leaves usually curl up and fall off.

Remedy

Organic
- Try simply removing affected leaves because rust develops late in the season, it doesn't always affect every leaf, and it's a pity to get rid of the entire plant (and lose the flowers). A spray of neem oil or manure tea *before* rust is noticeable might inhibit the rust. Leave plenty of room around plants and water late in the day, avoiding the leaves. Grow hollyhocks at the back of the flower bed, where you can't see the rusty bits. But the best antidote is prevention: pick disease-resistant varieties.

Chemical
- Use a lime sulfur spray before buds open.

Root Rot

Just about any plant can get root rot, but it's particularly prevalent during a cold, wet spring. (Roof rot is also a problem with houseplants that have been overwatered.) The symptoms are wilting and leaves turning yellow at the base of the plant. Also, an affected plant will just sit there in the ground, not growing. Dig it up. If roots are mushy, black, easily broken, and maybe smelly, that's root rot.

Remedy

Organic
- Try cutting off the affected bits with scissors. Add some more organic matter to the hole and replant. Don't overwater. Many plants recover from root rot, but if the weather continues cold and wet, they may not. If there's no improvement, dig them up again and destroy them. Avoid putting the same kind of plant in that spot again because the fungus that causes root rot stays in the soil.

Chemical
- None

WAYS TO KEEP CRITTERS AWAY

Once cats and dogs were the only four-footed pests that gardeners had to contend with. Now we have a whole menagerie out there. Wild animals keep invading urban areas because the pickings are better there than in the country and, as they have virtually no predators, the critters tend to stay and multiply. Whether we like it or not, learning to live with animal incursions is part of modern gardening. Get a dog if you are persistently bothered by animal pests and let it out on garden patrol regularly because canine protectors are often the best deterrent. Here are some other tips.

An old wire cupboard shelf makes a good animal deterrent.

Cats

To stop felines from digging, stick bamboo barbecue skewers in flower beds, close together. Cats don't like their sharp, pointy ends. Orange and lemon peel scattered about may repel them, as cats dislike anything that smells of citrus. An annual plant called *Coleus canina* (usually sold under the brand name Scaredy Cat) is widely touted to keep kitties away because they don't like its smell. But it has a strong scent only when the plant's leaves dry in late summer and you need a lot of them to get cats to pay attention. (In northern climates, it's also necessary to replace this plant every year because it's tropical.)

Bear in mind, too, that cats like bare, open spaces where the soil is dry and easy to scratch in. If you pack your garden with plants placed closely together, and try to keep the soil moist, feline trespassers will be less inclined to treat it like a public toilet.

Deer

While deer are undeniably beautiful animals, they can be hugely destructive.

Long lists of plants that deer are supposed to dislike frequently pop up in gardening magazines, but if they are hungry, these "rats with antlers" will try dining on virtually anything. Deer strip bark and the tender inner green layers off trees, killing them. They bite off the tops of plants and shrubs, then

Cats can be a nuisance in gardens, but they help to keep squirrels away. They will also catch mice, moles, and voles.

often spit them out. And they can flatten everything, trampling through flower and vegetable beds. Here are some tips:

- Deer don't have big teeth, so they tend not to like chomping on plants that have rough, bristly leaves (like squash or black-eyed Susans) or prickly leaves (like hollies *Ilex* or Scottish thistle *Onopordon acanthium*). They also avoid plants with pungent smells (like *Geranium maccrorhizum*) and anything that exudes an irritating sap (like *Euphorbia polychroma* or *Euphorbia myrsinites*). Plant these repellent plants as barricades, encircling or in front of your desirable plants. They are all easy to grow.

- Deer adore little shoots of, and the bark of, burning bush *Euonymus alatus*, sand cherry *Prunus*, young cut-leafed sumacs, and other leafy shrubs like forsythia and *Weigelas*. Do not bother to plant these if you live in an area with lots of deer! They will constantly tear at them in the winter months when they are desperate for something to eat.

- Deer can jump very high, so most fences won't keep them out. Barricades must be up to 12 ft (3.75 m) high, and electric fencing is often the only option.

- Deer don't like entering areas where they feel trapped. Position two long 4-ft (1.25-m) high fences only a few feet apart, and plant a skinny vegetable garden within that space. Deer will avoid jumping inside because it's too narrow for them to escape quickly. Prickly bushes (like blackberries) and branches of thorny buckthorn placed around vegetables may keep them out too.

- Fishing line and old CDs make a good poor man's version of a fence. Stick stakes 8 ft (2.5 m) long in the ground (cheap furring from a lumber yard is good, so long as it's solidly placed). The stakes need to be about 4 ft (1.25 m) apart. Then stretch four rows of fishing line along this fence, winding it around the stakes and pulling it taut. Hang CDs along the line. The CDs look like huge eyes at night, especially when there's a moon, and they will scare off the deer.

- Some people use human hair and scraps of strong-smelling soaps as repellents, hung in trees. Commercial repellants contain everything from coyote pee to sulfur. These sometimes work, but they smell disgusting and have to be constantly renewed.

Groundhogs and Gophers

Both are burrowing rodents whose tunnel systems underground are so complex they can cause embankments and garden berms to collapse. Both animals are very territorial and, once settled in a certain area, tend to stay. Gophers dine mostly on roots of plants, but groundhogs make a beeline for vegetable gardens, particular carrots, anything green and leafy, and beets. Traps are the best way to eliminate them. Tipping used cat litter repeatedly down their holes may persuade them to move on. With groundhogs, always look for two holes—an exit and an entrance—and dump the cat litter in both.

Mice

Mice girdle trees by feeding on the bark, and also tunnel through lawns. However, it's virtually impossible to banish them without resorting to traps, toxic poisons, or fencing buried in the ground. Hardware cloth, sunk 2 ft (.5 m) deep, may keep them out of flower beds and lawns, but you have to make sure they aren't already in residence, which can be difficult. The best deterrent is to get a cat or two, or even three. Female cats tend to make the best mousers.

Moles and Voles

We usually lump these two together, but they're different animals, causing different problems. Moles are carnivores, feeding on beetles, grubs, ants, and lawn grubs. They don't eat plants, and can in fact be useful to have around but their drawback is that they tunnel everywhere, causing mounds of excavated soil in lawns, flower beds, and pathways. Voles—sometimes called field or meadow mice—are vegetarians, feeding on many plants, but particularly seeds, bulbs, and rhizomes. Like mice, they're fond of girdling trees. Owls feed on voles, but as these rodents breed prolifically, cats make the best deterrent. You can also use traps, plunged into their tunnels, but removing a squished vole or mole from one of these metal gadgets is not for the fainthearted.

> **ALERT!**
> *Don't use powdered cayenne pepper on bulbs*
>
> While we'd all like to banish the bushytails, this method is cruel because it can blind squirrels and cause them incredible pain. Humane societies now frown on the practice of using powdered cayenne pepper. Hot pepper wax is preferable because it sticks to the plant and doesn't come off.

Pigeons and Starlings

Seldom a problem for ground-level gardeners, these avian pests may drive high-rise dwellers crazy, using balconies as toilets and kicking up a racket. Try stretching narrow mesh black plastic netting over the entire balcony opening, attaching it to walls and ceiling. (This works only in high-rises that have balconies inset into walls.) A kid's wire Slinky toy stretched along the balcony railing or outside-facing half wall will stop them from landing. So does shiny plastic string, or strips of flapping fluorescent tape. Artificial owls generally don't work unless they're the kind that have bobbing heads.

Rabbits

Cottontails are said to prefer young, tender seedlings, but, like deer, they will munch on anything when hungry. A local rabbit population will constantly invade vegetable gardens, and the only way to keep them out is to install close-mesh chicken wire fence all around the garden, sunk 6 in (15.25 cm) into the ground. Cats and dogs chase rabbits, however, and if they manage to snare a baby rabbit, shut your ears because rabbits make a horrible screaming sound. Losing one of the family will sometimes persuade Mr. and Mrs. Cottontail and the rest of the family that it's time to move on. Rabbits are often a huge problem in winter, like deer.

Raccoons

These are smart, persistent scavengers who'll eat almost anything. They rip up lawns, looking for grubs (see page 141), rummage around in flower beds (wild ginger *Asarum canadense* is a favorite target), and love fresh fish from a pond. Raccoons also make a mess because they knock over garbage cans in search of snacks and like to wash what they eat. (They will dip their food into any convenient water source and then splatter muck everywhere.) Raccoons keep coming back to the same place (often on the same night in the week!) and can quickly learn how to turn door handles and pry open garbage cans. Fox urine, baby powder, or human hair scattered about may repel them. Live traps baited with peanut butter may work too. However, the only things that really scare off raccoons are lights or a garden hose equipped with sensors that turn on when the critters come within their range.

Critters can be a problem with container-grown plants. These pots were knocked over by a visiting raccoon.

Skunks

Treat in the same way as raccoons.

Squirrels

They are the biggest offenders in many gardens, mostly because they attack our beloved tulips. When planting the bulbs in fall, don't add bonemeal because it attracts the bushytails. See page 66 and try these measures:

- Wrap bulbs in chicken wire or wide-mesh plastic netting.
- Give bulbs a suit of armor by encasing them in old coffee cans with wire netting or hardware cloth over the top. Remove the top

and bottom of the can before planting. Punch holes in the bottom of the can and sink it into the hole around the bulb.
- Smooth out the planted area. Squirrels are more likely to rummage around in soil that looks disturbed.
- Put human hair or prickly bits of holly in the tulip holes.
- Stick bamboo barbecue skewers closely together in the soil where you've planted. (This deters cats too.)
- Plant a skunky-smelling bulb called *Fritillaria imperialis* crown imperial. For more on this bulb, see page 120.
- If all else fails, plant daffodils instead. They're poisonous to squirrels, although some will still dig them up out of curiosity.

When tulips appear in spring, spray the buds with a garlic and water solution. Repeat after it rains. Try sprinkling blood meal around (Be warned: This will attract cats and wildlife like raccoons and skunks.)

Putting protective barricades over plants in their early growing stages is the best preventive measure against marauding wildlife.

Acknowledgments

Many thanks to Carol Cowan of the Netherlands Flowerbulb Information Center and Valerie Stensson of Sheridan Nurseries for their help with photographs. I also wish to express my thanks to my spouse Barrie Murdock, who took care of the technical stuff and also got roped into photographing various things in our garden to illustrate the book.

Photo credits:

All photos by Sonia Day and Barrie Murdock © apart from:
Pages 21, 26, 49 Sheridan Nurseries
Page 31 Netherlands Flowerbulb Information Center

Bibliography

FURTHER READING

In preparation for writing this book, I did a lot of reading. The following books are useful not only for their in-depth advice, but also the readable way in which they present the information.

The Well-Tended Perennial Garden: Planting & Pruning Techniques by Tracy DiSabato-Aust. Published by Timber Press.
Essential Gardening Techniques; U.K. Editor in Chief Christopher Brickell. Published by the Royal Horticultural Society.
Flower Gardening 1-2-3. Published by Home Depot and sold in Home Depot stores.
The Vegetable Gardener's Bible by Edward C. Smith. Published by Storey Books.
Mark Cullen's Ontario Gardening by Mark Cullen. Published by Penguin Books.
Marjorie Harris' Favorite Garden Tips by Marjorie Harris. Published by HarperCollins.

I also recommend a wonderfully informative book called *Everything Sold in Garden Centres (Except the Plants)* by Steve Ettlinger. Published by MacMillan in 1990 (now, unfortunately, out of print).

OTHER USEFUL SOURCES OF INFORMATION

On Hardiness Zones

United States: U.S. National Arboretum
http://www.usna.usda.gov/Hardzone

Canada: Agriculture & Agri-Food Canada
http://sis.agr.ca/cansis/nsdb/climate/hardiness/intro.html

On Climate Change and Gardening

Royal Horticultural Society, U.K.
http://www.rhs.org.uk/news/climatechange.asp

On Eco-gardening and Botanical Insecticides

Cornell University Department of Horticulture
http://www.gardening.cornell.edu/ecogardening

On All Pesticides

EXTOXNET, a complete database on pesticides and their effects maintained in a cooperative effort by University of California (Davis), Oregon State University, Michigan State University, Cornell University, and the University of Idaho: http://extoxnet.orst.edu/

Index

A
Acetic acid, 89, 130
Acidanthera, 70
Aegopodium
    podagraria, 23
    variegata, 23
African daisies, 84
Agricultural urea, 64
Alcea rosea, 98, 104
Alchemilla mollis, 23, 38, 104
Alliums, 55, 69, 92
Amaryllis, 32
American bittersweet, 38
Anchusa, 104
Annuals, 30-31
    buying, 28-30, 31, 114
    planting, 62, 66, 67
Antidessicants, 55
Ants, 134-35
Aphids, 114, 117, 118, 134, 135, 150
Artemisia, 38, 85, 100, 104
    ludoviciana, 38
Asarum canadense, 23
Asclepias tuberosa, 85, 93
Asperula odorata, 39
Asters, 98, 100
Azaleas, 55

B
Bacterial wilt, 151
Balcony gardening, 15, 69
Barrenwort, 22
Basil, 62, 94, 99, 108, 120
Bearded iris, 93
Bee balm, 93
Begonias, 32, 34, 70
Bergenia, 23, 100
Bigroot geranium, 22
Bindweed, 86, 91
Biological controls, 132-33
Bishop's weed, 23
Black peat ("black muck"), 46
Black spot, 151-52
Black walnut, 24-25
Black-eyed Susans, 19, 38, 94, 96, 98, 102, 107
Blanket flower, 94
Blight, 152
Blood meal, 66, 161
Blue bugloss, 104
Bonemeal, 66, 71
Borers, 136
Boston ivy, 38
Botrytis blight, 153
Brunnera macrophylla, 23
Bulbs, 31, 32
    buying, 32-34
    planting, 69-72
    spring-flowering, 31, 55, 63, 92
    summer-flowering (tropical), 31-32, 70
Butterfly weed, 85, 93

C
Caladiums, 70
Calcium, 52
California poppies, 85
Calla lilies, 70
Campanula, 93, 102
    rapunculoides, 38
Campsis grandiflora, 39
Cannas, 32, 34, 70
Catalogs, 34-35, 36
Caterpillars, 136-37
Cats, 156
Cayenne pepper, 72, 121, 127, 159
Cedars, 69
Celastrus
    orbiculatus, 38
    scandens, 38
Celosia, 62
Cercis canadensis 'Forest Pansy', 15-16
Chinese bittersweet vine, 38
Chinese trumpet creeper, 39
Chives, 94, 95, 104, 119
Chlorosis, 149-50
Chrysanthemums, 96, 99
City gardening, 15, 19, 71, 85, 87
Clay pots, 109
Clematis, 69
Cocoa bean fiber, 76
Coir, 48, 76
Coleus, 23, 62, 95
Coleus canina, 156

Colorado blue spruce, 69
Companion planting, 119-21
Compost, 46-48, 76
Coneflower, purple, 38, 39, 94, 98, 102, 105, 107
Container gardening, 31, 55, 70, 72-74, 108
Convallaria majalis, 22
Coreopsis, 85, 100, 107
Corn gluten, 55, 90-91, 130
Coronilla varia, 38
Cosmos, 62, 85, 93, 95, 98, 107
Creeping bellflowers, 38
Crocus, 69
Cross-border buying, 36
Crown imperial, 121
Crown vetch, 38
Cucumbers, 62
Cutworms, 137-38

D
Daffodils, 63
Dahlias, 32, 34, 70, 98
Dame's rocket, 38, 105
Damping-off disease, 113, 150, 152-53
Dandelions, 86, 87
Daylilies, 93, 103
Deadheading, 92-94
Deadnettle, 22
Deer, 156-58
Deleafing, 100
Delphiniums, 98, 100
Dendranthema, 96
Dianthus deltoides, 104
Disease-resistant plants, 113-14
Diseases, 149, 151-55
  prevention, 113-14
  treatment, 115-16
Dividing plants, 100-03
Dodecatheon, 22
Dolomitic limestone, 56
Downy mildew, 153-54
Drought-tolerant plants, 85

E
Earwigs, 120, 138
Echinacea, 38, 39, 94, 98, 105
Eleagnus angustifolia, 24
Epimedium rubra, 22

Epsom salts, 56
Escholzia californica, 85
Eucomis, 70
Eupatorium
  fistulosum, 98
  purpureum, 38
Euphorbia, 38, 85
Evening primrose, 38
Evergreen shrubs, 55, 69

F
Fertilizer, 49-55
  bulb, 71
  chemical, 53, 54-55
  numbers on bags of, 51
  organic, 53-54, 66
  slow-release, 66
Fish emulsion, 66
Flea beetles, 138-39
Foam flower, 22
Forget-me-not, perennial, 105
Fragaria, 38
Fritillaria, 69
  imperialis, 120-21
  meleagris, 121
Frost, 16-17, 62-63, 114
Fuchsias, 93
Fungal diseases, 60, 100, 149, 150
Fungicides, 125-26, 131-32
Fusarium wilt. *See* Damping-off disease

G
Gaillardia, 85, 94
Galium odoratum, 22
Garden centers, 14, 27-30, 115
Gardener's garters, 39
Garden
  maintenance, 81-82, 85, 87-88, 106
  in new subdivisions, 42
  planning, 18, 59-60, 65-66, 115
Garlic, 119-20, 121
Geranium macrorrhizum, 22, 100, 120
Geraniums (pelargoniums), 62, 85, 93
Gifts of plants, 105-06
Gladioli, 70, 98

Gladiolus callianthus, 70
Global warming, 15, 18, 115
Gophers, 158
Goutweed, common, 23, 106
Grasshoppers, 139
Gray mold, 153
Green lacewings, 117
Ground covers, 89
Groundhogs, 158
Grubs, 132

H
Hardiness zones, 13-15, 16
  map, 12
  "zonal denial," 15-16
Heather, 44
Hedges, 69
Hellebores, 100
Hemerocallis, 93
Hens and chicks, 85
Herbicides, 89-90, 124-25, 130-31
Herbs, 62, 94, 95, 104, 107-08, 120
Hesperis matronalis, 38, 105
Heucheras, 93
Himalayan balsam, 105
Hollies, 55
Hollyhock mallows, 105
Hollyhocks, 98, 100, 104, 107, 113, 154
Honeydew, 147
Hoses, 83, 84
Hostas, 23, 93, 102, 144
Hot pepper wax, 127, 159
Houseplants, 20, 55, 109-10, 142, 143, 146, 154
Hyacinths, 33

I
Impatiens, 23, 62, 94, 107
  glandulifera, 105
Insect traps, 133
Insecticides, 121, 123-24, 127-30
Insects, 28, 115
  beneficial, 117-19
  prevention, 114, 119, 120
  treatment, 115, 116, 123-24, 127-30, 132
Integrated Pest Management (IPM), 116
Invasive plants, 36-39

Iris, 93, 103

J
Japanese beetles, 139-40
Japanese knotweed, 38
Japanese spurge, 23
Joe Pye weed, 38, 98
Juglans nigra, 24-25
Juglone, 24
June beetles, 141

L
Ladybugs, 117-18
Lady's mantle, 23, 38, 104
Lambs' ears, 85
Lamium maculatum, 22
Landscaping fabric, 77
Lantana, 37
Lavender, 85, 107
Lawn care, 55, 116, 132
Leaf miners, 140-41
Leaf polish, 56
Leafhoppers, 140
Leaves, yellowing of, 149-50
Liatris, 85
Ligularia dentata, 84
Lily leaf beetle, 141-42
Lily of the valley, 22
Lime, 43
Loam, 41, 42
Lungwort, 23
Lysimachia, 38, 105
Lythrum salicaria, 39

M
Macleaya cordata, 39
Magnesium, 52, 56
Maiden pinks, 104
Mail-order companies, 34-36
Malva alcea, 105
Manure, 46, 47
    tea, 57
Marigolds (*Calendula*), 62, 85, 119
Mealybugs, 142
Mice, 158-59
Michaelmas daisies, 96
Microbia decussata, 69
Microclimates, 15
Milk thistle, 105

Mints, 38, 94, 95
Moles, 159
Molluscicides, 126
Monarda didyma, 93
Mugo pine, 69
Mulch, 67, 75-79
Mycorrhizal fungi, 56, 66
Myosotis, 105

N
Narcissus, 55, 69, 92
    dwarf, 23
Nasturtiums, 95, 120
Native plants, 17
Nematodes, 118, 142-43
Nicotiana, 62, 93
    sylvestris, 107
Nitrogen, 50, 51
NPK ratio, fertilizer, 51, 52

O
Oak leaves, 76
Obedient plant, 38
Oenothera, 38
    Missouriensis, 105
Onions, 119
Onopordon acanthium, 105
Ornamental grasses, 85, 96, 103, 108
Ornamental strawberry, 38
Osteospermum, 84
Ozark sundrops, 105

P
Pachysandra, 100
    terminalis, 23
Pansies, 62, 95, 99
Papaver somniferum, 105
Parasitic wasps, 118
Parsley, 108
Parthenocissus tricuspidata, 38
Peace lily, 20
Peat moss, 45-46
Penstemon
    barbatus, 93
    'Husker Red', 107
Peonies, 94, 97, 98, 100, 103, 107
Perennial plants, 30, 93, 95
    buying, 28-30, 114
    dry shade, 22-23

    fall cleanup, 107
    planting, 61, 66, 67, 114
Periwinkle, 22, 38, 100, 102
Perlite, 49
Perovskia atripicifolia, 99
Pesticides, 115, 121-22
    chemical, 123-26
    organic, 126-32
Petunias, 62, 93, 95, 99, 107
pH levels, 43-44
Phalaris arundinacea var. 'Picta', 39, 96, 108
Pheromone traps, 133
Phlox, 95, 98, 100, 113
Phosphorus, 50, 51
Physostegia, 38
Phytophthora blight, 153
Picea pungens var. *glauca*, 69
Pigeons, 159
Pinching plants, 67, 99
Pine needles, 44, 77
Pineapple lily, 70
Pinus cembra, 69
Plant disorders, signs of, 149-50
Plant growth, causes of poor, 18, 51
Plant tags, 14, 18, 19, 20-21, 60, 66, 101
Planting, 59-61
    annuals and perennials, 64-67, 114
    bulbs, 69-72
    in containers, 72-74
    roses, 74
    timing of, 61-62
    under trees, 21-24, 67-68
Plume poppy, 39
Polygonatum, 23
Polygonum aubertii, 39
Polygonum cuspidatum, 38
Poppies, 94, 98, 105
Portulaca, 85, 95
Potassium, 50, 51
Potatoes, 62
Potting soil, 72
Powdery mildew, 153-54
Praying mantis, 118-19
Pulmonaria, 23
Purple coneflower, 38, 39, 94, 98, 102, 105, 107

Purple loosestrife, 39

R
Rabbits, 159-60
Raccoons, 160
Rain barrel, 84
Reed canary grass, 39
Rhododendrons, 44, 55, 153
Rhubarb, 120
Ribbon grass, 39, 96, 108
Road salt, 64
Rock gardens, 111
Rooftop gardening, 15
Root rot, 155
"Rose blindness," 110-11
"Rose food," 54
Roses, 56
    diseases affecting, 113, 114, 151, 152
    hybrid tea, 74
    in cold climates, 74, 108-09
    rugosa, 74
    watering, 83, 151-52
Roundup, 89-90
Rudbeckia fulgida, 19, 38, 94, 98
Russian olive, 24
Russian sage, 99
Rust, 154-55

S
Sage, 107
    Russian, 99
Salvias, 93
Sand, sharp, 49
Scale, 143
Scillas, 69, 92
Scottish thistle, 105
Sedums, 85, 99, 100, 107
Sempervivum, 85
Shade, 19-20
    dry, 21-22
    plants, 18, 22-24
Shasta daisies, 93, 97, 98, 102
Shearing plants, 94-96
Shooting star, 22
Siberian cypress, 69
Sidewalks, gardens next to, 44
Silver fleece vine, 39
Silver lace vine, 39

Silybum marianum, 105
Skunks, 160
Slugs, 92, 100, 106, 119, 120, 126, 143-45
Snails, 126, 145
Snapdragons, 62, 93, 99
Snowdrops, 23
Soaker hose, 84
Soil, 41-42, 60
    improving, 44-49
    kits for testing, 43
    pH of, 43-44
    potting, 72
Solarization, 91
Solomon's seal, 23
Sowbugs, 145-46
Spathiphyllum wallissii, 20
Sphagnum moss, 45
Sphagnum peat moss, 46
Spider mites, 146
Spined soldier bug, 119
Spring cleanup, 79, 110
Sprinklers, 84
Squash, 62, 152
Squirrels, 72, 120-21, 160-61
Stachys lanata, 85
Staking plants, 96-99
Starlings, 159
Stone mulch, 77
Straw, 77, 79
Sulfur, 52
    horticultural, 44
Sunflowers, 85, 94, 98, 107
Sunlight, 18-19, 20-21
Sweet woodruff, 22, 39
Swiss stone pine, 69

T
Tagetes, 62, 85, 119
    tenufolia 'Lemon Gem', 119
    tenufolia 'Tangerine Gem', 119
Thinning plants, 100
Thrips, 146-47
Tiarella cordifolia, 22
Toads, 134
Tomato hornworms, 147-48
Tomatoes, 56, 62, 97, 98, 99, 113, 152
Tools, 66, 71, 82, 87, 92, 95, 103
Topsoil, 45

Trees
    leaves of, as mulch, 76, 107
    mulching, 75
    planting under, 21-24, 66, 67-68
Trumpet vine, 39
Tuberous begonias, 24
Tulips, 55, 63, 69, 92, 120, 121

V
Variegated plants, 25
Vegetables, 62, 120, 152, 155
Verbascum, 105, 107
Verbenas, 93, 95
Vermiculite, 49
Veronicas, 93
Vinca minor, 22, 38, 100
Vinegar, horticultural, 89, 130
Vines, 108
Viola odorata, 23, 105
Violets, common, 23, 105
Voles, 159
"Volunteer" plants, 94, 103-05

W
Walls, planting beside, 22
Watering, 67, 83-85, 110
Weeding, 85, 87-88, 106
Weeds, 86-87
    remedies for, 88-92, 111
Whiteflies, 120, 148
Wild cucumber vine, 86
Wild ginger, 23
Wilting, 84, 151
Wind damage, 68-69
Winter
    preparing garden for, 55, 96, 106-10
    shoveling snow near plants, 64
Wood chips, as mulch, 77
Wormwood, 85

Y
Yarrows, 85, 93

Z
Zinnias, 85, 95, 98
Zones. *See* Hardiness zones